# Sacred Cows
and Hot Potatoes

Annual Policy Review 1991/92
National Center for Food and Agricultural Policy

# Sacred Cows and Hot Potatoes

## Agrarian Myths in Agricultural Policy

*William P. Browne*

*Jerry R. Skees*

*Louis E. Swanson*

*Paul B. Thompson*

*Laurian J. Unnevehr*

*With Forewords by Bob Bergland and Rudy Boschwitz*

Routledge
Taylor & Francis Group

LONDON AND NEW YORK

First published 1992 by Westview Press, Inc.

Published 2019 by Routledge
52 Vanderbilt Avenue, New York, NY 10017
2 Park Square, Milton Park, Abingdon, Oxon OX14 4RN

*Routledge is an imprint of the Taylor & Francis Group, an informa business*

Copyright © 1992 Taylor & Francis

Library of Congress Cataloging-in-Publication Data
Sacred cows and hot potatoes : agrarian myths in agricultural policy
  / by William P. Browne . . . [et al.].
    p.  cm.
  Includes bibliographical references.
  ISBN 0-8133-8557-1 — ISBN 0-8133-8558-X (pb)
  1. Agriculture and state—United States.  2. Agriculture—Economic aspects—United States. I. Browne, William Paul, 1945–
HD1761.S17 1992
338.1′873—dc20                                                          92-5155
                                                                           CIP

ISBN 13: 978-0-367-28652-1 (hbk)
ISBN 13: 978-0-367-30198-9 (pbk)

# Contents

# Illustrations

## Figures

## Boxes

## Tables

# Foreword

## Bob Bergland

It was in March 1977. I had just been appointed Secretary of Agriculture by President Jimmy Carter and was in Kansas at a meeting of wheat growers. I met with a mostly younger set of growers who brought their grievances. In 1973–74 wheat prices had doubled in eighteen months because of poor crops in the world and some political realigning in the USSR. The growers believed they had heard government officials say that the so-called "farm income problems" had been solved. During that same period, land prices started climbing and had doubled in about four years. By 1976, the whole world had responded to the price increases, production grew, and prices collapsed.

These younger people had bought or rented land during that price run-up period, and by 1977 with the collapse of wheat prices, they had been left out to dry. Their plea was that I should raise price-support loan rates to a level of $5 a bushel so that they could service their land debt. That request reinforced what I had been told by an economic mentor in Minnesota, Dr. Willard Cochrane, who argued that farm profits are always capitalized in land values.

Later that same year, I was in Mississippi at a meeting of about two hundred black farmers who were struggling to make a living on small acreages. One of their pleas was that I raise cotton loan rates to a level that would keep them in business. From that meeting, I went to California's famed Central Valley and became acquainted with a gentleman who farmed more than 100,000 acres of cotton. The stark reality between a group of struggling blacks and this California megafarm dramatized to me the impossibility of stabilizing that group of small-scale black farmers through cotton loan rates without unduly enriching that large California ranch.

These experiences were very difficult for me emotionally, because I had grown up in a prairie populist environment dedicated to the concept of family farming maintained by high price supports. I found that my basic doctrine didn't hold up in the real world.

This book deals with these conflicts in a sharp and incisive way. Having a market-oriented farm program posed a dilemma when, in the early 1970s, wrong market signals were sent to these Kansas wheat growers or when poor beef prices caused the wholesale selling off of cattle. The plowing up of thousands of acres of grassland and the planting of soybeans for $10 a bushel was an environmental disaster. And everyone, including the farmers, knew it. These examples underscore the need for public policy that tempers market prices by interventions when carefully planned public needs are at risk.

In 1979 I directed the first-ever comprehensive USDA study, "A Time to Choose," on the structure of farming in the United States. The study documented the enormous concentration in production, especially of perishables. It also documented the importance of nonfarm jobs to more than a million families, classified as farmers by the Census, who farmed on a very small scale but lived comfortably on income from nonfarm jobs. The traditional diversified family farm was being squeezed out by the large, heavily integrated forces on one side and by the growing number of families who had the economic security of jobs on the other.

This study is a forceful argument for fresh thinking in the next round of general legislative debate, which will start in 1992. The book is a "must read" for everyone who has an interest in the subject—which should be all of us.

# Foreword

## Rudy Boschwitz

The strength of this book is that it defines American agriculture as it really is, not as it exists in popular mythology, and then proceeds to show the incongruity between current government programs and those realities. The authors demand change, and I agree. This book illustrates and proves—perhaps better than any other I've seen—that U.S. agricultural policy has not kept pace with the changes that have taken place worldwide, on American farms, and in the rural areas of our country.

While in the U.S. Senate, I recognized many of those things as well, and in 1985, along with Senator David Boren (Democrat of Oklahoma), I introduced the Boschwitz-Boren bill as an attempt to redress some of the conflicts between agricultural policy and current realities in agriculture. Our bill "decoupled" production from support payments so that markets, not farm programs, would influence what farmers planted. The bill challenged existing programs as a threat to the ability of U.S. farmers to compete in international markets. It challenged the effectiveness of commodity programs to do what they claimed—save family farms and rural communities. And it recognized the increasing burden on taxpayers of continuing to transfer wealth to some who were already wealthy. The bill had bipartisan support—but unfortunately not enough.

This book provides a concise and compelling basis for updating our outmoded perceptions and beliefs about U.S. agriculture. It suggests a framework for redirecting our thinking about agricultural policy and the goals we wish to accomplish with it. It points out with clarity the many ways that agrarian myths temper our beliefs about agriculture and thus have made policy change an almost impossible process. The list of reasons why government finds it difficult to reform agricultural policy and make it do what the rhetoric says it does is insightful.

An interesting recent calculation, found in the *National Journal* (September 14, 1991), should give policymakers some pause. Entitled "The Counterproductivity of Farm Subsidies," the article shows that those countries with the most intrusive and expansive agricultural subsidies, allegedly designed to "save the family farm," have the fastest declining agricultural sectors. Those that intrude the least are the only countries that have an expanding agriculture. They must have gotten an advance copy of this book!

What a challenge that should be to us. America is blessed with huge expanses of fertile soil in a temperate climate, and through the vast agricultural heartland is a wonderful labyrinth of rivers providing cheap transportation to the world's markets. Given a chance and equal access to markets—a level playing field—American farmers will prosper.

And so will rural America, where the world's most remarkable agricultural infrastructure gives us a large, added advantage. If American agriculture were allowed to produce for markets and if the infrastructure were used to full advantage, rural America would certainly prosper.

The authors, appropriately for a book of this kind, do not offer policy prescriptions. But their conclusion—that the burden of proof that policy is in the public interest should shift from those who want to effect change to those who want to maintain the status quo—is certainly right, if difficult to accomplish.

A copy of this book should not only be on every policymaker's shelf, it should show visible evidence of having been well read.

# Acknowledgments

Few authors owe as much as we do to our mentor, George "Ed" Rossmiller. As director of the National Center for Food and Agricultural Policy (NCFAP), he has provided intellectual leadership to all of us in our policy-related work. But beyond Ed's help, we would like to acknowledge that NCFAP, as an organizational entity, gave each of us opportunities, research support, a climate for intellectual growth, and great friends in what is, in our opinion, one of the W. K. Kellogg Foundation's wisest investments. This book, which is NCFAP's 1991/92 Annual Policy Review, is a direct result of Ed's efforts to assemble an interdisciplinary team to address the role of myths in policymaking. We owe a debt to him for envisioning this project and also to James T. Bonnen and Robert L. Paarlberg, who helped to shape the book's original outline.

Although we are responsible for the contents, many others have helped us to produce, document, and refine our book. Courtney Harold, Jaime Casteneda, and Kathryn Kampmann at NCFAP provided able and thorough research assistance. Two colleagues, Gerald C. Nelson and Barry Barnett, furnished valuable comments on early drafts. And finally, our editor, Sheila Ryan, and Wheeler Arts turned our document into a readable and attractive text.

In addition to the intellectual support provided by NCFAP and our colleagues, we are grateful for NCFAP's logistical and financial support. We are also grateful to the W. K. Kellogg Foundation for partial support of the production of the manuscript. Release time provided by the Ford Foundation helped one of our authors to pursue this project and to gain insights on Capitol Hill.

*William P. Browne*
*Jerry R. Skees*
*Louis E. Swanson*
*Paul B. Thompson*
*Laurian J. Unnevehr*

# 1 An Overview

This is a book about the lack of accountability in American farm policy. Things are not going as they should. Public programs are not appropriate to current social needs. Agricultural policy, we find, responds to the past far more than it does to any current issues and problems. There are two primary reasons for this.

First, policymakers and the public are held captives of the agrarian past through myths that have evolved over time. Agrarian myths are what philosopher and legendary agricultural administrator John M. Brewster called "first principles," which we use to understand our nation's agriculture. These myths rest on a nearly blind faith that agricultural interests will both want and do the right things.

But alas, faith is easily misplaced in a world ordered by politics. Farmers, as we show throughout our discussion, have faced tremendous obstacles simply trying to stay in business. There is no pastoral wonderland. Technology, markets, and the political economy have intervened and changed the entire structure of agriculture. So, as a second reason for clinging to the past, agricultural interests have promoted our attachment to the "first principles" of agrarianism. Farmers and society are portrayed as under constant attack by those who would disrupt stability. Sometimes the bogeyman is subsidized foreign competition; just as often he is the unsympathetic U.S. consumer. Public policies must, no matter who the assailant, be conjured up to protect the sector.

Indeed we find that U.S. public policies are so protective of domestic agriculture that they contribute little to the public well-being. While agricultural policies could prove useful to the general good, the current and massive array of rules and programs serves only narrow interests by often outmoded means.

That conclusion is not ours alone. The five of us are hardly that original; and we are too much a part of agriculture's own establishment to take potshots at farm programs. No. Contemporary U.S. agricultural policy has been under assault for years. Its practitioners have countered substantial challenges: in the 1980s, farmers mobilized in grass-roots organizations to raise prices; during that same period, agricultural economists galvanized their efforts to argue in public forums for less intervention in price policy. Political ideologues have had their go as well. Prairie populists of the left, such as Marty Strange, have written extensively about policy failure. Think tank conservatives on the right, such as James Bovard, have leveled many of the same charges. Common to farm, academic, and theorist critiques is the contention we share: agricultural policy has lost its relevance.

Agricultural policy has pressures other than protest that prompt reform. The federal budget's mounting deficit has already led to substantial cuts for traditional farm programs. In international affairs, a reconstructed view of the world order has subjected agriculture to change everywhere. Europe 1992 will bring a truly common market; the breakdown of the Communist Block in Eastern Europe promises new uncertainties in world markets; and the emergence of new exporters among less developed countries brings intense competition from trade partners. A domestic-international policy nexus also occurs as the costs of agricultural programs prove collectively too much for both the national budgets and the trade advantages of many countries. In a logical sense, the United States and the world should face an era of agricultural policy reform.

*We feel that reform might never come, even though it makes obvious sense.*

We, however, feel that reform might never come, even though it makes obvious sense. That worry prompts this cooperative venture and the book that our multidisciplinary team has written through our often intense personal discussions. In the ten chapters that follow, we will walk our readers through a morass of myths associated with agricultural policy.

The format of the book is straightforward. Each chapter stands on its own, although all are linked to form an integrated whole. Chapter 2 explains the basis of agrarian myths, their importance to our country, and how they are odd mixes of reality, outdated tradition, and fundamental values.

The next eight chapters address prevailing myths, ones that most need to be critically reexamined in their policy impact. All, as first principles, are cornerstones of misunderstanding that perpetuate public policy debacles. In each case, we outline several assumptions that frame a specific myth, move on to debunk the outmoded particulars, and then attempt to salvage important values associated with the original. In the last stage, which is a "rebunking" or regeneration of the old myths, we urge society to hold onto those elements of its traditional values that still lend currency and relevancy as policy guidelines. Some aspects of each myth, we find, still contain important truths.

Chapter 3 is about the confusion between farm policy and rural policy. The two must be seen as distinct, even though they are often confused. In Chapter 4, we turn more specifically to understanding farm policy. There we caution readers about the fallacy of relying on the image of the average farm.

Warnings continue as we turn next to farm income and market problems. Never confuse production and productivity, we caution in Chapter 5. Chapter 6 then goes on to show that farm price supports are not effective in stabilizing farmers' income. In Chapter 7, we demonstrate how U.S. agriculture depends on the world economy. Our intent throughout these three chapters is to refute the tendencies in policy debates to simplify economic conditions.

Chapters 8 and 9 turn to what Don Paarlberg referred to in 1980 as "new agenda" agricultural issues. Specifically, we deal with the environment and food. The intervening years should have brought adaptation to these concerns. But policy debates about the environment and food needs still descend to name calling. As we explain, the notion persists that (1) farm programs can easily be made compatible with these issues and (2) farmers can take care of all problems associated with them.

Lest we be accused in our critique of just being advocates of big government and regulation, Chapter 10 takes on our final myth. Never assume, we argue, that government programs do what they claim. Sometimes they cannot work their desired ends. At other times, those who put them into play simply lack the incentives to let them.

With these myths in mind, the final chapter sets forth an action agenda of what we want for agricultural policy. Taken and expanded from our concluding remarks in each of the earlier chapters, our discussion centers on one unifying theme. Specifically, we conclude that agriculture is important and that the United States needs an agricultural policy, but only one where its policy advocates can prove their own case. The burden of proof for making agricultural policy decisions should be shifted. In the past, the burden of justifying policy content has fallen on those who wish to change it. When advocates of reform could not satisfy decision makers, old programs persisted, and they continued to be passed off as serving traditional goals. The defenders, with these rules in motion, have an advantage. We ask only that agricultural interests pick up the mantle of proof. If they want programs to continue, they should demonstrate beyond some reasonable doubt that the public well-being is served.

In developing this theme to our conclusion, we must explain one final point. We hope to spotlight policy myths and their implications because, as noted earlier, the public is not well served if they go unchallenged. Thus, for us as authors and critics, *the public is our constituency.*

As we first planned this project, we thought otherwise. We looked for select targets to whom we could direct this effort. Was it *urban policymakers* who needed more information about falsehoods they encountered? Or was it *rural policymakers* who needed to be reminded of the mythic content of the rhetoric they employ? Or was our target to be *representatives of the agricultural establishment* at large, and especially land-grant college specialists, who may have unwittingly fallen victim to idealistic interpretations of their own historic mission? Because agricultural organizations are grappling with a sense of their future mission, critical exploration of why they do what they do is undoubtedly needed.

We finally recognized that any selection of special constituents was wrong because it was not they who are aggrieved. While we hope that those groups noted above read and use this book, it is not for them or other select groups that we wrestled with these ideas and laboriously wrote them down. We hope only that readers think of the public well-being in their policy responses and become somewhat more public regarding. We have picked up that perhaps overly idealistic charge as our challenge in preparing the following pages. Why?

As a recent Kettering Foundation survey by Richard Harwood discovered, citizens are angry rather than apathetic about politics. They see themselves as victims of political and communications processes that do not truly inform them about issues. They want to know more, or at least they want that opportunity. Citizens, the Kettering study finds, want more than superficial, three-minute sound bites that barely cover policy issues and problems and explanations of why policymakers respond as they do. Citizens want more, are angry about not getting more, and do indeed deserve more from those paid through the public trust. That is what we attempt to give them here.

# 2 Never Assume That Agrarian Values Are Simple

Perceptions are indeed reality. Resting upon our values and ideological position, our perceptions help us to make sense of agricultural policy issues. But these very ways of seeing what we think is reality also cloud our vision.

As a society, we pass on perceptions and values from generation to generation. Opinions about social issues depend, in part, on these inherited values. In many regards, socialization through inherited values makes a tremendous amount of sense. People simplify things by accepting the values and perceptions of previous generations. Socialization has obvious positive attributes. A common framework of perceptions makes daily social interaction among people possible, even when they do not know each other. As society and the economy change, however, what might have once been good common sense can become outdated. And outdated ideas can become the basis for legitimizing policies that may no longer fit. It is important that citizens challenge their perceptions and assumptions and the evidence presented about values. This book challenges many of the *agrarian myths* that our society has passed on since the time of Thomas Jefferson. In the process, we will attempt to "rebunk"—that is rehabilitate—these key ideas that shape perceptions of agriculture, farmers, and rural society. We do not intend to destroy these myths, but rather to recast them for understanding *current* agricultural issues.

First, we must recognize that beliefs and values motivate and shape public responses to social problems. Beliefs and values influence policy. Many authors, including James T. Bonnen and William P. Browne, show how agrarian myths frustrate attempts to reform farm policy. In the context of this book, *The American Heritage Dictionary* gives a highly relevant definition of *myth*: a notion based more on tradition or convenience than on fact. *Myth* also refers to widely shared stories about how society ought to be organized. This type of myth depends less on what is perceived to be real and more on what we want to be real. For our purposes, we will tend towards the former definition by demonstrating how specific social and economic interests have used agrarian myths to gain legitimacy in the eyes of a general public.

## Agrarian Myths

In a democratic society where citizens' beliefs and values influence policy, it is important to challenge these beliefs and values. They have taken on too much prescriptive meaning based on outlived conditions. Given an infinitely complex world, people often do one of two things when they attempt to choose between public policy alternatives. They may defer to inherited beliefs and values, sometimes leading them to make choices based on rather naïve criteria. This pattern can create problems when times change. Or people may defer to someone else. We often forgo our rights to influence an issue when others seem more knowledgeable or more directly affected. We seldom offer alternatives to ideas suggested by those in authority. This pattern empowers those who have a significant interest in the issue. In the case of agricultural policy, those outside the agricultural establishment may choose not to become involved. Both of these patterns reinforce traditional perceptions and norms, the first simply by relying upon them uncritically, the second by neglecting to take up issues where traditional ideas are likely to be challenged.

*Broadly, agrarian myths involve beliefs about the relationship between agriculture and democracy.*

This chapter traces the origins of many of our beliefs about farmers and agriculture. They form a complex web of many interrelated ideals. Broadly, agrarian myths involve beliefs about the relationship between agriculture and democracy. Some of the myths derive from Thomas Jefferson's well-known aphorisms about farming. But the central agrarian myth is not all of Jefferson's making. Parts of it are rooted in 19th century attitudes toward nature and human potential. Still other parts are founded upon religious ideals that were carried to the New World by Protestant sects hoping to escape persecution. Each of these sources of philosophical inspiration for agrarian values was a phenomenon of unique historical forces at work in 18th and 19th century America. Although these historical events produced enduring philosophical insights, we need to understand how these insights depended upon historical circumstances for their initial application to agriculture.

## How the American Agrarian Myth Was Created

Thomas Jefferson's thoughts on American farms are among the most frequently quoted political values in American history. In *Notes on the State of Virginia*, Jefferson (p. 290) says that "those who labour in the earth" have (if anyone has) been chosen to receive God's "peculiar deposit for substantial and genuine virtue." In his 1785 letter to John Jay, Jefferson (p. 818) writes:

> Cultivators of the earth are the most valuable citizens. They are the most vigorous, the most independent, the most virtuous, and they are tied to their country and wedded to its liberty and interests by the most lasting bonds.

Untold numbers of politicians and political commentators have quoted these passages to bring forward the idea of a political duty to preserve and protect farms (Box 2.1). But Jefferson's words are open to other interpretations. While they have a political message of preserving the farm basis of the American economy, they also and more aptly reconcile self-interest with the public good.

The quoted passages, it must be noted, were written between the Declaration of Independence and the Constitution, which established the United States as a republic in 1789. The constitutional debate between Jefferson and the Federalists, notably Alexander Hamilton, centered on whether power should be concentrated or broadly distributed. A reason for concentrating power lies in the individual citizen's tendency to shirk public responsibility. Most of us press for private interests. The public good is hard to see.

Federalists accepted the idea that some people were in a better position than other people to recognize the public good. In England, for example, an educated class of gentlemen had financial interests that were tied to large landed estates. Federalists thought that estate holders would more closely identify their interests with those of the state than would manufacturers, merchants, and tradesmen. The latter interests were seen as more naturally selfish than those of farmers, simply because of the nature of their investment. People in predominantly urban occupations can convert wealth into capital and abandon a crumbling government; planters cannot. Thus the grower is the more reliable citizen.

Even now this view holds. Contemporary poet and essayist Wendell Berry, for example, strongly emphasizes the way that Jefferson reconciled self-interest with community and with the necessity of stewardship. Because farmers must stay in one spot, they must learn to get along with their neighbors and take an interest in long-term social stability. The virtues of honesty, integrity, and charity that

Box 2.1.

## Jefferson and Citizenship Ideals

It is useful to think of citizenship as a role that we prepare for, as an actor studies a script. The founding fathers clearly believed that some people would be far better equipped than others for the part. A person's ability to play the role of citizen hangs on having sympathy with the part, on being able to motivate oneself to faithfully carry out the citizen's responsibilities. Performing as a citizen thus requires a certain kind of character. Jefferson argued that, in America, freeholding yeomen farmers would have that character because their farming would make it obvious that the state's ability to protect their lands served personal interests. Because they could not remove themselves (or their farms), they had to accept the need for resolving political differences in ways that did not endanger the stability of the state.

The cited passages from *Notes on the State of Virginia* and the letter to John Jay mentioned in the text are part of Jefferson's reply to the Federalists. Jefferson notes how it is not the aristocrat but the common farmer who occupies the land in America. His remarks start from the idea that landowners, unlike other people, make a stronger identification between self-interest and the common good and hence are more virtuous *as citizens*. A landed population is therefore less apt to be tempted by short-term, unsustainable government policies. In America, however, the land was owned by small, yeoman farmers. Jefferson's argument is a rationale for aligning political power with a previously existing pattern of land holdings. In the American context of the age, the argument favored democracy and undercut key premises of the antidemocratic themes of the Federalists.

Jefferson's statements about farms must be read as part of his strategy to place government power in pursuit of the common good. Small farms were important to Jeffersonian democracy because they promised temporary relief from the threat of mob rule. Jefferson knew that the structure of society would eventually change. But he hoped that, if it could be established on a democratic basis, new traditions would emerge to serve the public good.

For Jefferson, farmers were valuable as citizens because they would be well placed to see the convergence of personal interest and the public good. The idea of "public good" at work here is the stability and fiscal responsibility of the society. The ability to understand the link between public and private interest is a mark of citizenship.

promote community are the same virtues that promote the farmer's own interest.

For this reason, a nation of farmers was thought to be stronger than a nation of traders or manufacturers. A manufacturer is not so firmly tied to a community and hence is less firmly wedded to the virtues Berry finds so important. Businesspeople spoil the air, exploit the local workforce, poison the wells, and then pick up their assets and move on down the road when the business environment becomes hostile. Although Jefferson (not surprisingly in his era) does not take up environmental themes explicitly, Berry's interpretation shows how Jefferson's concern for the long-run public good can be given, by implication and updating, an environmental twist.

Jefferson's argument was intended to suggest that farming would induce long-running patterns of conduct that would be broadly characteristic of American society as a whole. Given the experiences of his day, Jefferson had good reason to think that a nation of pragmatic farmers would be easier to govern than a nation of merchants, manufacturers, and landless laborers.

By the 19th century, the role of citizen-farmer was even richer and more romantic than it ever was in Jefferson's time. The romantic side of agrarian values has at least two themes. First is the notion that nature is a formative element in the American national character. Second is the related idea that hard physical labor is a prerequisite for achieving the virtues necessary for self-realization. Both of these themes were commonly espoused in 19th century America.

Ralph Waldo Emerson and Henry David Thoreau were the intellectual leaders in the 19th century movement that redefined human purpose as communion with nature. Emerson in particular thought that farm work was an especially good example of the morally successful life (Box 2.2). Emerson's thought is therefore a crucial link between Jefferson and more recent agrarians such as Berry. Farmers apply the work ethic in a natural setting. By doing so, they become a model for how everyone *should* live, at least in an idealized world.

Emerson's works develop the link between farming and character, just as Jefferson's stress the link between farming and citizenship. Emerson implies a triangle of farming, citizenship, and character by providing a plausible way of understanding how farming contributes to achieving moral virtue. The good farmer, as the good person, follows the path of the true poet, aiming toward authentic realization of self. Most will not reach the poet's vantage point on nature, and no one can remain there indefinitely. According to this romantic vision, a democracy of farmers provides a land rich in signs that point the way toward self-realization. Such a farmer-owned agriculture brings together a community of individuals who actualize human potential

Box 2.2.

## Emerson and Nature

Emerson's writings on nature were explicitly linked to his concept of moral character. Achieving one's potential for appreciation of value and self-expression was the overriding purpose of human life. For Emerson, the fullest realization of human possibility was in poetic creation, and appreciating nature was a vital part of the creative process. In his essay *Nature*, Emerson writes that the human spirit has been fitted for the sights and experiences of the natural world. The artificial world of the city forces people to work in an environment that is not conducive to the realization of natural powers. The city is a world of broken time and shrouded light, a place where the human impulse to bestow meaning upon the commonplace is likely to select temporary or debauched objects of attention. While apparently simpler than city life, the world of nature is, for Emerson, a place of infinite detail and richness, well suited to the human impulse for inquiry, representation, and experimentation.

Unlike Thoreau, Emerson saw nature as a place where human beings were busily at work transforming the natural landscape. Nature was not merely an object of contemplation. It was the arena where the human drama was enacted. Nature provided the material that must be transformed if one is to grow in the realization of self, which is mankind's purpose on earth. But nature must be transformed in ways that are authentic expressions of human spirit, that ring true to the innermost core of being. This problem is most acute for the artist or poet who aspires to create, but who lacks the vision and instincts to realize a true work of art. Aiming to please the critic, the false poet creates not from a strongly felt grasp of place, but by imitating forms that have won the approval of others. Compared with false poets, farmers more closely realize the human ideal, despite a relative lack of formal education. Farmers care little for the opinions of others. They must create in their own life a work of art whose only judge is nature herself. If nature judges the work well, farmers thrive; if not, they suffer. Given that they work in and with nature, farmers develop traits of observation and invention that are tailored to humanity's evolved capacities of sensation and ratiocination. For this reason also, the farmer's life activity is a far more authentic production of the creative spirit than is the false poet's imitation of art (Corrington). The farmer is thus a signal pointing the way to spiritual fulfillment for even the true poet, who will be better served by observing life than by studying great books.

far more fully than any other society in human history, including ancient Greece. While Emerson rarely discusses citizenship directly, his thought illuminates the underlying value of democratic citizenship for the fulfillment of human purpose.

### Agrarian Ideas Today

In more recent times, the Jeffersonian argument and its romanticized version are both recognizable in populist writings by Harold Breimyer, Wendell Berry, and Jim Hightower. But there is a major shift in meaning. While Jefferson and Emerson selected farmers as *embodying moral and political ideals* that should be applied to *all* citizens, contemporary populists use these arguments *as reasons for exempting farmers* from criteria routinely applied to others. In recent debates, the populist themes have emerged in opposition to those who have questioned whether farm policies promote the most efficient investment of human and financial capital in American agriculture. The efficiency argument has promoted the idea that those who leave farming during periods of crisis, or those who *would* leave if commodity support programs are suspended, simply represent an appropriate reallocation of society's resources to nonfarm activities. It seems clear enough that efficient allocation of society's resources was the criterion applied when U.S. heavy industry began declining in the 1970s. Why should farms be any different?

Populists present two unique strands of reasoning for making an exception for family farmers. Both strands are philosophically conservative, although they have been used to promote seemingly liberal policies that resist market forces. The first strand of reasoning sees farms as an important "safety valve" necessary to preserve individual liberties in capitalist societies. The second strand sees family farms as repositories for family values and hence for traditional ways of defining personal loyalties within a framework of community.

Breimyer describes the role of farming as an entry-level occupation that serves as a "safety valve" for the American economy, although Hightower has been its most active political proponent. Behind their work is the assumption that the American concept of liberty has always included an assurance that no one would be forced to work for wages in order to live. American society would guarantee its citizens an opportunity to "be their own boss," although neither success nor great wealth is assured by this right. The small family farm has institutionalized this right throughout American history. Those with little formal training who did not want to work in factories could take up farming.

But transformations in agriculture, especially in the costs of entering farming, have now all but eliminated the right to this important opportunity. If the populists are to be believed, as the right to self-employment disappears, the political legitimacy of American capitalism goes along with it. The populist quarrel is not with capitalism, but with the relative position of small farmers to capitalist institutions or any large bureaucratic institution, such as the U.S. Department of Agriculture (USDA) or the land-grant universities.

This traditionalist version of populism has been argued by Berry. In his view, small farms are good because they cultivate virtue in the farm family. The traditionalism in Berry's approach can be understood by considering the typical Jeffersonian era household. According to social historian Ruth Cowan, *the husband* traditionally looked after the household, cared for and tended the land, deriving *his* title from *his* house to which *he* was bonded. The housewife and the husband worked the land, hence the term *husbandry* for what we would now call farming. Their economic security depended upon working together and "husbanding" their resources. The success of the household depended upon both the man and the woman successfully completing a diverse set of well-defined tasks that were thoroughly interrelated by gender role. Cowan (p. 25) writes:

> Buttermaking required that someone had cared for the cows (and . . . this was customarily men's work), and that someone had either made or purchased a churn. Breadmaking required that someone had cared for the wheat (men's work) as well as the barley (men's work) that was one of the ingredients of the beer (women's work) that yielded the yeast that caused the bread to rise. . . . Women nursed and coddled infants; but men made the cradles and mowed the hay that, as straw, filled and refilled the tickings that the infants lay on. Women scrubbed the floors, but men made the lye with which they did it.

Before the industrial revolution, the American farm household usually required gender-specific roles to produce a wide range of commodities and services. These relationships were mediated by a paternalistic culture. The farm family thus represented a closed social system in which self-reliance required people to interact with others of the opposite sex and hence to carry out fundamentally different social roles. In his 1977 book *The Unsettling of America*, Berry offers a long chapter entitled "The Body and the Earth." There he argues that households now chiefly need cash income—a need that can be met only by working outside the household. As the household has become identified with the consumption of goods and services, sexual behavior has itself become increasingly understood as a consumption activity. Men and women perceive each other as mutual possessions, rather than as productive partners with their own tasks.

Berry's thought revives some ancient and traditionalist ideas of value. Moral obligations issue forth from roles that unify economic and political status. Duties are grounded in loyalties to particular people who are bound to one another in time and place. Virtue is found in living up to the role one inherits by being situated in a community or a family. The particularity of being situated means that men's roles differ from women's, that adults' roles differ from children's, that farmers' roles differ from cobblers', blacksmiths', and carpenters', and that each person must fulfill the requirements of his or her role. Those who do so are virtuous; those who are overcome by jealousy, competitiveness, greed, and other vices fall short in the moral quest. Such is an order, agrarians argue with passion, that must be defended and, given the unwanted changes of industrial societies, even rekindled.

## Agrarianism, Myth, and Public Policy

Agrarian populist arguments have thus been applied to public policy as reasons to preserve family farms or, more frequently, to preserve farming as a way of life in general. Yet it is far from clear that the values advocated by Breimyer and Berry are truly served by programs that protect farmers from economic forces. Furthermore, seldom do escapees from the urban factory retreat to a well-integrated farm family, or at least so few that they should not be used as the basis for supporting farm programs. As far as Breimyer and Berry may be from the realities of American farming, the original aims of Jefferson and Emerson are even farther. In contemporary farm policy debates, farms are rarely promoted as models for all citizens.

Even if we doubt the substantive vision of Berry or Hightower, the family farm is certainly a deeply significant part of American history. It provided the backbone of American economic and cultural life for more than two hundred years. Even if the agrarian vision is no longer valid, it was clearly valid once. Even if we do not have to be farmers to learn independence and self-reliance, the heritage of the family farm is a reminder of these virtues. The family farm is a theme in children's literature, in films and novels, and in cultural life as a whole. It thus continues to serve in symbol what it once served in fact (Figure 2.1).

It is far from obvious how mythic values ought to be accommodated in policy. The farms of the agrarian ideal form a part of their cultural heritage for many Americans. The agrarian myths may be vitally important to our feeling of community and common purpose, and a failure to address them seriously could haunt society's attempts to address agricultural policy. As it stands, they are contested symbols, vague images of how agriculture ought to be, or once was. Their lack of specificity means that competing political interests can easily appropriate them.

When farm policy advocates say that farming is a way of life worthy of public support, they draw upon an image of farming from times past. But how misleading is their claim? The farming way of life was important because it was thought to encourage character traits that were essential for a well-functioning democracy. It is hard for us to see how farming can bear the burden of preparing us for our role as citizens in postmodern America, yet the myths of the American rural past may still in some way serve us in a future where fewer than 2 percent of Americans will live on farms. The citizenship, community, hard work, stewardship, and self-reliance found in farming certainly must continue to be values widely shared with the rest of society if democracy is to succeed.

There is irony in the fact that an agrarian philosophy that stresses the importance of community and the public good should be used to promote the interests of a few. Contemporary agriculture is a far cry from Jefferson's and Emerson's vision. It is neither possible nor appropriate that we revert to the technology and social organization of the 19th century. What is needed is a way to rehabilitate the philosophical foundations of the agrarian myth. Those foundations imbue agriculture with values and responsibilities that make it an integral part of moral and political life. What is important in Jefferson is that public policy—indeed the Constitution itself—interacts with the life patterns of Americans. Policy should not be evaluated solely in terms of how benefits are distributed, or who wins and who loses. Rather, it should be evaluated in terms of how the broader patterns of conduct do or do not reinforce the development of character traits that will serve to moderate the flaws of democratic government.

Jefferson's vision did not fail in his own time, nor has it failed in ours. That two centuries of Americans have been willing to assume public responsibility is evidence of his wisdom. Indeed his vision may have succeeded too well: we tend to think that democratic procedures will always work in the end. We have accepted the belief that markets, elections, and political processes work in the abstract. Do we need public-spirited citizens who possess values formed by hard work, stewardship, and self-reliance? The fear of losing these values, despite our lack of understanding the need for them, is certainly strong. Perhaps the agrarian myth—and its extension to a broader society, translated through children's literature, through books and films, through the study of history and the arts—has been enough to moderate the tendencies of the mob. But if it has, can we rely on the agrarian myth indefinitely? Probably not.

*Today the public seems less willing to see the farming community as the principal source of moral inspiration and virtue.*

Today the public seems less willing to see the farming community as the principal source of moral inspiration and virtue. Some city dwellers now see farmers as glorified welfare recipients or as willful polluters, rather than as paragons of virtue. While the old images are false, these new images are also false. Farmers face unique difficulties, and it is in the public interest that policies be constructed to resolve farm problems. Urbanites have reason to suspect those who portray farmers as heroes, however. Cynical uses of agrarian myths to promote private interests, whether by agribusinesses or farm interest groups, will further erode the cultural foundations of agrarian ideals. Farming is different today, but agrarian ideals, ideals focused on the public good, are still relevant to farming and to agricultural policy. Our book will take up a series of specific myths, or public policy assumptions, where the influence of a larger more encompassing central agrarian myth continues to be felt. We will examine each one, its derivation, and the current reality of American agriculture. The intent is not to show that they are false, but to show how, despite changes, a kernel of truth still exists in each. It is those kernels of truth that form the basis for rebuilding American agricultural policy.

# 3 Never Confuse Farming with Rural America

Nowhere is the influence of 19th century romanticism more evident than in linking farm policy to rural social policy. The romantic image assumed that farming encompassed all rural life. In this vision, country life is commonly portrayed as a bucolic landscape full of hard-working, middle-class farmers. Pictures of rolling hills, carefully crafted rows of corn and soybeans, pure white homesteads and neat barns beside blue silos readily come to mind. No doubt, one has seen these images hanging on family room walls. The picture does *not* include a mobile home park or the towering smokestacks of a factory surrounded by chain-link fences.

This bucolic image is so powerful within our collective psyche that it is routinely used in advertisements to sell everything from soft drinks to insurance to chemical products. Rural America is seen as the last vestige of the most desirable part of U.S. history. This rural America of our dreams persists because it is wrapped up in our desire for ties to the land, economic independence, and community support.

The problem is that the image fails to reflect reality, at least beyond that of our perceptions. But in life, as opposed to politics, perceptions are not necessarily reality. The trailer park and the factory are very real. Moreover, life inside them is typical for far more rural people than is life on that picture-perfect farm. Farm policy proponents who suggest otherwise are merely harking back to some other era. Today only a minority of rural people rely on farming. But a great proportion of farm people rely on off-farm income, and most of these jobs are not in agriculture.

*This does not mean that farming is unimportant to rural well-being.* Most farm families make up the more wealthy segment of rural society, so their well-being is of consequence. The point, however, is that most rural people have no direct attachment to farming. And farm policy has a negligible impact on their lives. Government support for farmers is therefore no surrogate for rural development except in a very few areas of the country. Even so, the belief that rural well-being is synonymous with farm well-being continues to be a powerful myth that undermines alternative rural development policies and their funding.

## The Power of Myth

In the second chapter, myths were shown to be outmoded and traditional notions that help sustain fundamental social beliefs. They are full of inaccuracies because of how they are used and how they evolve. Because Americans value farming and rural regions, myths help preserve both heritages. But by confusing farming with rural, the myth prevents adjustments to new social and economic problems of the countryside.

*Confusion happens when federal farm programs are sold as sound economic policy for all rural regions.*

Such confusion happens when federal farm programs are sold as sound economic policy for all rural regions. Rural Iowa and some other midwestern states may come close to embodying this linkage, but most of rural America does not. Just as there is little point in discussing the *average farm* (see the next chapter), so there is little evidence for the image of an *ideal rural community*. Rural America is composed of highly diverse economies and cultures. Textile mill towns in the Carolina piedmont do not look like the energy boom and bust towns of the Southwest, timber communities of the Northwest, or the farming communities of the Midwest. Rural America is not a culturally homogeneous ocean surrounding the nation's culturally diverse metropolitan islands. Yet federal rural development policy assumes that it is.

Nevertheless, most myths do spring from real world experiences. Indeed rural America once was dominated by farming, as well as by forestry, mining, and fishing. This myth has been further strengthened by a common sense assumption about the well-being of any local society. The dominant economic activity of a rural area and the social organization of that economic activity will undoubtedly have the greatest impact on people's lives.

But this axiom applies to all forms of rural economic activities, not just to farming. Mining, forestry, and fishing exerted equally strong influences on their local societies. Appalachian company towns and the textile company towns of the Southeast are good examples. These communities have seen poverty and a lack of democracy because of the relationships between owners and workers. But there is a difference in the way farming communities have been treated. Mining, forestry, and fishing have never been culturally cherished industries. Rural life has never been seen as dependent on them, although this is clearly inaccurate for Appalachia, the northern Great Lakes, and coastal regions.

## Economic Activities That Dominate Rural America

In most areas, farming does not dominate rural America as it once did. In 1790, the year of the first census, the United States was primarily agricultural. Even then, however, family farms were not the rule. Plantations were clearly dominant in the South. As recently as 1920, almost a third of the U.S. population was in farming. Among rural people, two-thirds were in farming. But the middle and late decades of the 20th century have witnessed a remarkable change in how and where rural people make their living. Farming, manufacturing, and service industries have all been affected.

The *great transformation* of rural America, as we shall see in later pages, is one of two very real, not mythic, stories. The first tells of the transition from earning a living by extracting natural resources (agriculture, forestry, mining, and fishing) to working mostly in manufacturing, services, and government. The second story tells of a change in class and social structure. Most of the two thirds of rural Americans working on farms at the time of the 1920 census owned some part of the farm. They were largely their own bosses. Others owned small retail establishments that catered to the needs of farmers. Only a minority of rural people, most notably in mining regions, worked for someone else. Today most rural people work for someone else. The same is true for many farm people.

### The Extent of Change in Farming

Each of the extractive industries has steadily declined in the number of workers and units. While technological changes made it possible to displace workers with machines and other innovations, technology alone did not cause the transformation. Changing economic forces were also causes. After all, technology does not lead farmers and mine owners to maximize profits. Rather, they adopted new technologies to compete successfully with others. In farming, the big switch was to specialized commodity production. Farms that once supplied much of the material they used and processed significant portions of their crops now routinely grow only one or two commodities.

This trend accelerated after World War II as resources were freed to modernize farms more quickly. The United States experienced a sharp

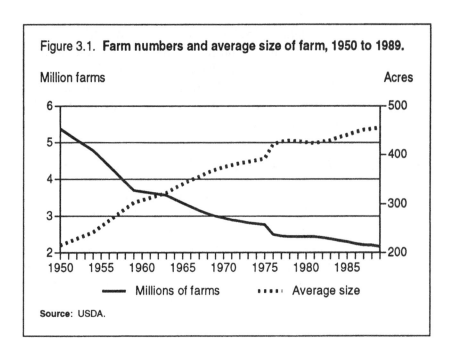

Figure 3.1. **Farm numbers and average size of farm, 1950 to 1989.**

decline in the number of farms, along with a rapid increase in average farm size. Both of these changes in farm structure can be seen in Figure 3.1. During this period the United States lost, on average, 2,000 farms a week. The great majority of these farms disappeared because the succeeding generation chose, for whatever reasons, not to stay in farming.

### Emergence of a Dual Farm Structure

While revealing, Figure 3.1 hides another profound change in U.S. farming: the emergence of a *dual farm structure*. Duality refers to the existence of essentially two farm sectors, one made up of very large farms and the other of small farms. Middle-sized farms are slowly disappearing. Nationally, before World War II, a majority of the wealth produced in farming was created by a large number of farms. Today only a small proportion of all U.S. farms produce most of farm-ing's wealth. While 70.7 percent of farms in 1988 had gross sales of less than $40,000, they accounted for less than 10 percent of sales (Figure 3.2). At the same time, 1.4 percent of farms with sales greater than $500,000 accounted for 36.6 percent of sales.

As Figure 3.3 shows, the movement toward a dual farm structure appears to have been steady over a considerable period of time. A look at the market influence of only the largest 5 percent of U.S. farms reveals that their market share expanded from 38.3 percent in 1939 to 54.5 percent in 1987, an increase of 16.2 percent in about fifty years. Three things are important. First, the United States has always had the

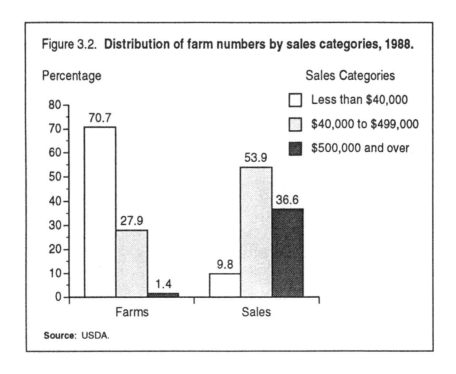

Figure 3.2. **Distribution of farm numbers by sales categories, 1988.**

Percentage

Sales Categories

□ Less than $40,000

▨ $40,000 to $499,000

■ $500,000 and over

Source: USDA.

makings of a dual farm structure. For example, in the South, yeoman farms coexisted with large plantations. Second, the degree of this duality has intensified significantly during the last two decades. Third, this polarization of the farm structure strongly implies that farm policies should be sensitive to fundamental long-term differences in the needs of both small and very large farms. More than thirty years ago, the Eisenhower Administration was perceptive enough to recognize this. But even they were not the first; their recommendations emphasized points made by the Country Life Commission of the Theodore Roosevelt era (Browne 1988).

### Increased Off-Farm Work

Another feature helps to explain farm structure: the steady increase in family members taking off-farm jobs. According to USDA, a slowly growing proportion of farm family income comes from these jobs. Nonfarm job growth has occurred for two reasons. First, as the size of farm necessary to secure a middle-class income increased, farm families had to expand their acreage. But many could not. They turned instead to off-farm work to supplement farm incomes. Off-farm jobs were also attractive because they provided a buffer against fluctuations in farm income. Second, the changing rural economy simply created more manufacturing and service jobs, which provided more opportunities to work elsewhere. In turn, off-farm employment became a reliable strategy for staying in farming, especially when

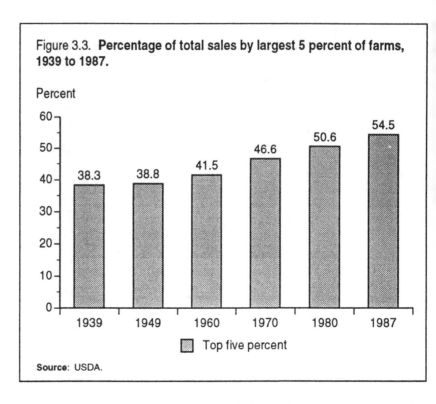

Figure 3.3. **Percentage of total sales by largest 5 percent of farms, 1939 to 1987.**

Percent

Source: USDA.

farmers were faced with unexpected sharp downturns, such as the 1980s farm depression.

The prevalence of off-farm jobs explains a lot about the duality of farm structure. Smaller farming operations tend to survive because of off-farm income. Data on the varying dependence of farms on off-farm income in 1988 are presented in Table 3.1. Clearly, farms with less than $40,000 needed off-farm income. An even sharper view of their need is provided in Figure 3.4, which uses data from Table 3.1. Ninety-three percent of total family income for farms with less than $40,000 in gross sales comes from off the farm. This compares with 22.5 percent for medium-sized farms and 3.5 percent for the very large farms. As Table 3.1 also shows, even medium and very large farms earn a substantial amount of money off-farm.

Table 3.1 demonstrates yet another aspect of farming, one that supports the notion that farming is important for rural well-being. Farm families earn, on average, higher incomes than do other rural residents. Even farms with less than $40,000 in sales averaged $28,422 in 1988. Medium and moderately large farms averaged $39,931 and $104,692. These figures should be accepted with caution, however. Many of the larger farms are operated by several families, and dollars often get spread thin. Moreover, about 20 percent of farm families live in poverty. Even so, the bulk of farm families are better off than most of their rural neighbors.

### Agriculture: More Than Just Farming

As industrialization and urbanization increased, some agricultural operations stayed on and near the farm while others did not. The raising of plants and animals obviously did. Production factors associated with farm equipment, fertilizers, specialized packaging, and the like moved off the farm into agribusinesses. Most of the jobs associated with business went not to town but to metropolitan areas. This shift is important to note because the number and type of agricultural nonfarm jobs has greatly expanded, while farming declined. Restaurants, supermarkets, and transportation were the growth industries of agriculture in the last half of the 20th century.

If agriculture is seen as all types of jobs associated with the processing and marketing of food, almost 19 percent of the U.S. labor force in 1988 was engaged in its work. As shown in Figure 3.5, most of these jobs are in metropolitan areas. Fewer than 20 percent of agricultural jobs are filled by the rural labor force.

## The Rural Economy

While farming was undergoing a fundamental restructuring, as were all of the extractive industries during the post-war period, the

---

Table 3.1.
**Distribution and Sources of Average Income by Sales Class, 1988**

| Sales Category | Number of Farms (1,000) | Percent of Farm Numbers | Percent of Cash Receipts | Household Cash Income, $ Net Cash | Household Cash Income, $ Off-farm | Household Cash Income, $ Total |
|---|---|---|---|---|---|---|
| Less than $40,000 | 1,555 | 70.8 | 9.8 | 1,988 | 26,434 | 28,422 |
| $40,000 – $99,999 | 320 | 14.5 | 13.6 | 25,252 | 14,679 | 39,931 |
| $100,000 – $499,999 | 293 | 13.3 | 40.0 | 87,402 | 17,290 | 104,692 |
| $500,000 and over | 30 | 1.4 | 36.6 | 762,830 | 27,891 | 790,721 |
| U.S. | 2,198 | 100.0 | 100.0 | 27,265 | 23,527 | 50,792 |

Source: USDA.

---

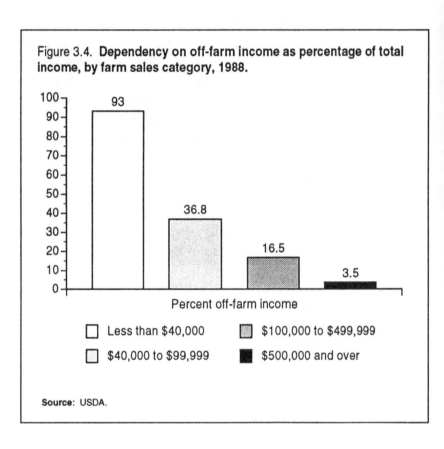

Figure 3.4. **Dependency on off-farm income as percentage of total income, by farm sales category, 1988.**

Percent off-farm income

☐ Less than $40,000          ▨ $100,000 to $499,999
☐ $40,000 to $99,999       ■ $500,000 and over

Source: USDA.

nonfarm rural economy was going through its own transition. Rural employment patterns were increasingly influenced by urban patterns. To a great extent this occurred because rural markets developed with improvements in secondary roads and bridges, completion of the interstate highway system, and the geographical expansion of metropolitan areas.

The South's experience illustrates the rapid change in employment. Sharecropping collapsed, migration opportunities for African Americans were poor, and the world market for goods manufactured in the region grew. Thus a minor industrial revolution took place during the 1960s. The rural South and some other regions became a haven for manufacturing and service industries on the periphery of the national economy. Job expansion was impressive, but it constituted primarily low-wage, limited-benefit opportunities. Consequently, it is not surprising that all of the persistently poor counties in the United States are nonmetropolitan. More than 90 percent of these are located in the South.

By the 1990s, nonmetropolitan employment was dominated by the service sector and manufacturing industries. Figure 3.6 provides a sense of how nonmetropolitan employment is divided among those

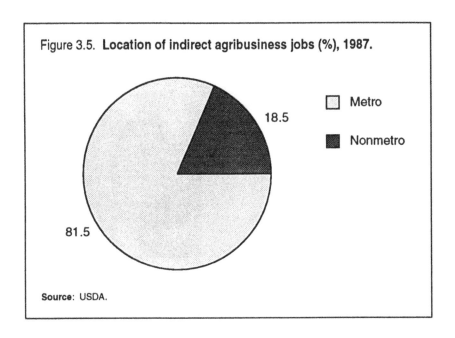

Figure 3.5. **Location of indirect agribusiness jobs (%), 1987.**

Metro

Nonmetro

18.5

81.5

Source: USDA.

who work. Almost 40 percent of all rural workers are employed in the service sector. Another 17.6 percent manufacture textiles and parts for core industries located primarily in metropolitan areas. Fewer than 10 percent work in agriculture. More, 16.8 percent altogether, work for government agencies, usually the local school system and highway department.

These percentages reflect the national economy, but with great variation locally. While rural areas nationally show great employment diversity, most local economies depend on only one or two key industries or employers. By 1986, only 516 nonmetropolitan counties were dependent on agriculture. An agriculturally dependent county counts only 20 percent of its earned income from farming. At the same time, 577 nonmetropolitan counties were dependent on manufacturing. And it is more difficult to be counted as a manufacturing county because at least 30 percent of income must come from this sector.

While 516 agriculture-dependent counties may seem impressive, these counties accounted for less than 7 percent of the entire U.S. nonmetropolitan population. Manufacturing-dependent counties accounted for 32 percent of the total nonmetropolitan population. However the employment pie is sliced, few rural people are directly or indirectly dependent on farming.

During the 1980s, employment in the nonextractive sectors expanded in rural areas, while all extractive industries suffered losses. As Kenneth Deavers notes, during the 1970s and 1980s each declined by 50 percent in its share of the rural labor force. Trends for several of

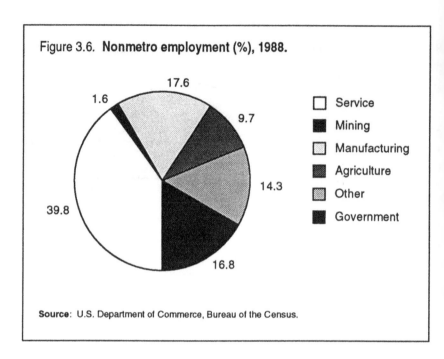

Figure 3.6. **Nonmetro employment (%), 1988.**

Legend:
- Service
- Mining
- Manufacturing
- Agriculture
- Other
- Government

**Source**: U.S. Department of Commerce, Bureau of the Census.

the key employment sectors are shown in Figure 3.7. The construction, service, and manufacturing industries led in nonmetropolitan job growth in the 1980s.

These major changes in the structure of rural employment were not accompanied by marked improvements in rural well-being. The level of poverty is similar to that of the 1960s. During the 1980s, non-metropolitan poverty actually exceeded that of metropolitan central city counties. Moreover, because most rural industries are on the margins of the national economy, these rural areas are afflicted earlier and longer during national economic recessions. Even farm investments of capital earnings overwhelmingly flow out of the rural locale to metropolitan centers of finance. Other evidence of a fraying rural social fabric can be found in the alarming rate of rural bank failures, environmental degradation as the cost of economic development projects, poor educational systems and low levels of education, and sub-standard health services.

Not only are rural areas less dependent on farming, but in most locales the new industries are not providing enough secure and high paying jobs to create or maintain quality communities. Rural poverty persists. But this does not mean that farming is unimportant for rural economies. Farming is still one of the few high income industries in rural areas. Rather, these findings mean that all forms of employment are important to local societies with few opportunities and little economic diversity. The point is that the dependency relationship for most rural areas has changed. And those communities are not likely

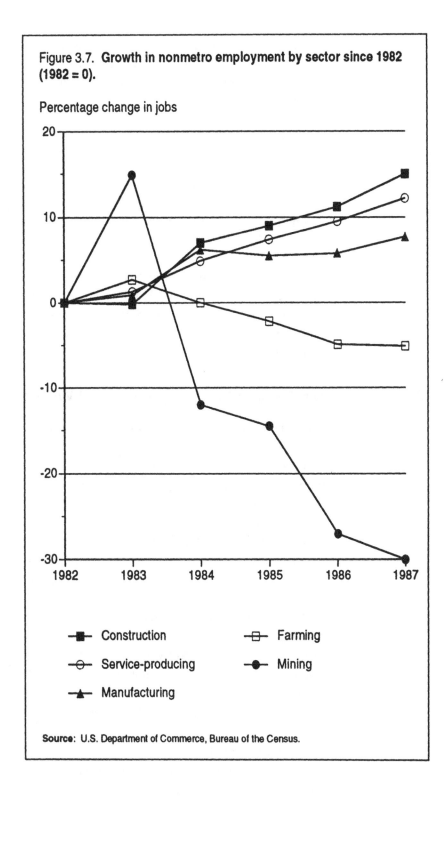

Figure 3.7. **Growth in nonmetro employment by sector since 1982 (1982 = 0).**

Percentage change in jobs

Legend:
- ■ Construction
- □ Farming
- ○ Service-producing
- ● Mining
- ▲ Manufacturing

Source: U.S. Department of Commerce, Bureau of the Census.

to reverse this unfortunate trend in the near future and return to a dependency on farming. This conviction is widely shared among rural development experts. The following quotation from Norman Reid (p. 12) captures this concern:

> A large percentage of adults with high school educations or less live in nonmetro areas. . . . These skills leave rural workers ill-prepared for the modern economy. Most new jobs created during the 1980's demanded higher levels of education. At the same time, little growth in new jobs occurred in rural areas at any education level. . . . The inability of the rural economy to add high skill jobs creates strong pressures for the educated to leave rural areas.

Rural labor forces have been greatly transformed. For farm families who depend on off-farm work, the choice is often between the hard place of small farming and the rock of low paying off-farm jobs. Unfortunately, thinking about the means for achieving rural well-being has lagged far behind understanding these changes.

## Altered Assumptions About Farming and Rural Well-Being

We should pay attention to this lag in critical thought and some of the misinterpretations neglect keeps alive. Changes in rural labor forces and rural class structures have only recently led to questions about whether rural economic well-being is dependent on farming. The prevailing view, as indicated in Figure 3.8, sees a one-way relationship. Farm well-being determines rural well-being. Based on the assumption that the dominant economic activity was once farming, this view has diverted attention from thinking about rural policy alternatives because it holds out farm price policy as the best solution through trickle down.

The call for policy alternatives does not mean that in places where farming is the primary source of employment it has no economic importance. On the contrary, it does. But the type of farm makes a difference. Family farms are thought to be far superior to corporate farms in their social consequences for communities. But defining a family farm or an industrial farm is open to interpretation (see Box 3.1).

Considerable evidence underscores the belief that communities dominated by industrial farm structures have low standards of living. Walter Goldschmidt's study of two California communities in 1944 and recent work by Dean MacCannell for the congressional Office of Technology Assessment (OTA) have supported this assumption over a forty-year period. But their research does not mean that small farms are necessarily better for rural communities.

Figure 3.8. **The traditional view of the relationship between farming and rural communities.**

Farm well-being ⟶ Rural community well-being

The OTA study for the South found a curvilinear relationship; that is, counties with either small-scale or large-scale farms had similar low standards of living. And mid-sized and presumably family and larger-than-family farms were found to be associated with much better standards of living. But does this amount to an overwhelming influence of farming in determining rural community well-being?

Except for the MacCannell study of industrial farm states, none of the remaining four OTA studies found evidence of a strong influence between a county's farm structure and various indicators of its social and economic well-being (see Box 3.2). Why? Two reasons are given. First and most obvious, as we have detailed, farming is no longer the dominant economic activity for most rural areas. Second, while there have been overwhelming declines in farm numbers and increases in farm scale, evidence of major qualitative changes in who owns the remaining farms is limited. Regions characterized by industrial farming operations have historically had this type of farming. The remaining farm structure in regions historically dominated by owner-operated farms continues to be so dominated.

Despite qualitative changes in technology, financing, federal programs, and markets, farming is still a household-based operation, although on a vastly larger scale and with more hired labor. The other expected qualitative change is the increasing dependence on off-farm income by small-sized and to a lesser extent medium-sized farms. This does not mean that historical trends will not eventually lead to industrial farms, but only that such an important shift has not yet occurred. And it may not.

Box 3.1.

## What Is a Family Farm?  A Corporate Farm?

Given the importance of symbols and images, disagreements in defining a family farm are not surprising.  We have no lack of reasonable definitions, but some people stand to gain politically by identifying their interest with family farms.  The most useful definition is offered by rural sociologists Kevin Goss, Richard Rodefeld, and Frederick Buttel.

They simply define a *family farm* as any operation where the family owns a majority of the capital resources (land, machinery, buildings), makes the majority of management decisions, and provides the majority of labor.  Operations where the family does not supply the majority of labor, but nonetheless works the farm, they term *larger-than-family farm*.  Similarly, an operation where the family does not own the majority of capital resources but does supply the majority of labor and management is called a *tenant farm*.  Today, tenant farms can be both quite large and profitable.  Both of these farm types, and the next, are based on household production, but with differing combinations of ownership and labor.

The sociologists' final category is *industrial farms*.  These farms are not based on any form of household production.  Rather, they are owned by one group of people, managed on a daily basis by another person or group, and worked by yet another group.  Just like most factories, there is a division of labor among the owners, managers, and laborers.  Industrial farms can cover a wide range of operations.  One of these is the *corporate farm*.  Part of a vertically integrated agribusiness interest, corporate farms account for less than 5 percent of all income generated by farms.  For the more typical industrial farm, a family owns the capital resources, but hires a manager, who in turn selects and directs hired laborers.  An increasingly common industrial farm is one where a farm management firm contracts with former farm kids who continue to own the land but do not wish to manage it themselves.  There is no evidence that corporate farms are a serious threat to household production, especially in regions historically dependent on family farming.

Most commercial farms are owned and operated by the same individual or family.  They vary considerably in the ownership of production resources and hired labor, but the operators live in the community and earn a living from farming.

## Dual Transformation of Rural Society

The class structure of an otherwise changing farm sector, whether industrial or household, has remained somewhat stable since World War II. But in the meantime, rural society has been transformed across two major dimensions. These are no less important and in many ways even more impressive than the transformations in farming, although they have not received nearly as much attention. The *dual transformation* of rural society's class structure is characterized by a decline among the other extractive industries (mining, timber, and fishing) and the growth of manufacturing and service sector jobs. Most rural people, including farm people with off-farm jobs, work for someone else. Rural labor forces now resemble urban labor forces, although they usually lack a corresponding proportion of high income jobs. This dual transformation has consequences for both farm structure and community well-being.

---

Box 3.2.

### Are Big Farms Bad for Rural Communities?

Whether or not *big* farms are bad for rural communities has been a hotly contested issue for some time. As is often the case with farming issues, it all depends. Both relatively small farms and very large industrial-type farms are associated with unfavorable indicators of rural community well-being. But this association is only partly due to scale. Some areas with small farms are not unfavorably influenced by their smaller scale. Tobacco-dependent counties, for instance, tend to benefit from the protection the tobacco program affords smaller producers. However, areas associated with the legacy of sharecropping, another form of very small farming, compare unfavorably.

Similarly, large farms have different influences on rural community well-being. Mega-farms with their industrial-type worker-manager-owner relationships depend on large, unskilled labor forces. They are consistently associated with very unfavorable indicators of rural well-being. However, large farms that are owner-operated but with large hired labor forces, for instance on midwestern dairy or corn and bean operations, do not seem to negatively influence rural communities.

Farm and community studies generally suggest that it is not so much the scale of operation but the social organization of the farm that influences rural communities. Large farms with industrial-type relationships tend to have negative influences, while owner-operator farms generally have positive influences.

---

This means that an alternative model of factors influencing rural well-being must be built into our thinking about public policy reform. This model, presented in Figure 3.9, continues to emphasize the primacy of a locale's dominant economic activity and its social organization. But it can also be applied to those rural areas where nonfarm economies dominate. This way of viewing farm and community relations recognizes the importance of farming, but places it within the context of the entire rural economy.

Where the nonfarm economy is dominant, its fortunes and the way it is organized, whether through industrial or household systems, will strongly influence a rural community's well-being. This model also points to the influence that the nonfarm economy has on farming. For farm families who depend on off-farm income, the well-being of the nonfarm economy can determine their standard of living.

This too is not a one-way relationship. The farm sector may also influence the nonfarm sector by providing a relatively high quality labor force, especially for industries that require less skill. Moreover, because farm families are unlikely to migrate when laid off during periods of economic recession, they provide a ready reserve of labor.

This model does not imply that either farm families or other rural families generally benefit from their dependency on the nonfarm economy. The earlier OTA reports found that neither manufacturing nor service industries universally promoted higher standards of living. In fact, in some regions such as the South, a greater dependency on these two sectors was often associated with lower standards of living. The dual transformation of rural society has seldom made rural areas better places to live.

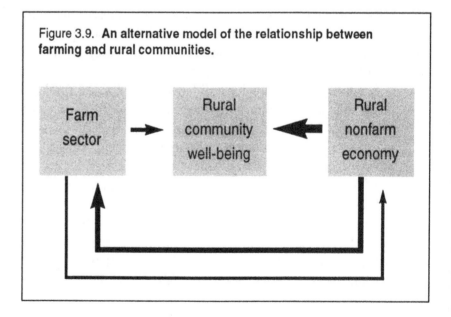

Figure 3.9. **An alternative model of the relationship between farming and rural communities.**

# Rebunking the Myth

As universal statements, claims that rural America's fortunes are tied to the well-being of the farm sector are incorrect. Hence farm policy cannot be sold as a rural development policy. Federal programs have aided farm concentration and the dual farm structure. So, in the aggregate, they have not helped family farmers or rural communities. But the part of the myth that says the dominant industry is important in shaping a rural area's social and economic well-being makes good sense.

We find little merit in totally accepting or in totally rejecting the myth. Farming has importance for rural well-being. As we noted earlier, farmers tend to be the wealthier members of rural communities, and they are often the primary source of property taxes. Their well-being counts, but in proportion to what they contribute to an area's total economy and labor force.

Despite the agrarian myth, nonfarm rural issues are national issues. With a quarter of the U.S. population living in nonmetropolitan areas, their economic and social conditions do have an impact on the nation's economy and standard of living. Moreover, rural areas account for about 90 percent of the nation's natural ecology. With rural poverty and the degradation of the rural environment comes a need for national policy.

## The Case for Rural Development

The present rural development policy has been referred to as minimalist, triage, and nonexistent (see Box 3.3). Concern for rural areas sometimes cycles with public concern for poverty. For example, the Great Society proposals in the 1960s included measures to address rural poverty. Unlike poverty programs, however, there has never been a systematic rural development program. Oddly, the farm financial crisis in the 1980s triggered a renewed interest in the plight of rural people. But it was a serendipitous discovery. The media, in search of the consequences of the farm crisis, found the same conditions that others had identified in the 1960s.

Altering the minimalist rural development policy of the past will be a formidable task. The dilemma, as Louis Swanson has pointed out, has at least five dimensions. First is the myth that the fortunes of the farm sector determine those of the rural area—hence farm policy is a viable surrogate for rural development. Second, social, economic, demographic, and other types of information on rural areas is woefully limited and even misleading.

Third, there is a pervasive assumption, very much associated with the current triage policy, that little can be done to help rural areas left

---

behind. The odd corollary is that these areas will not pose much of a problem for the nation, given their isolation from the national economy. Will their people, their problems, and their ability to touch the rest of society simply disappear?

Fourth, rural areas are often treated in isolation from larger regional, national, and international economies. The interconnectedness of rural issues with policy issues in other areas is frequently overlooked. For example, nonmetropolitan and metro central city counties are remarkably similar in terms of poverty, unemployment, crime, and other socioeconomic indicators.

A fifth dilemma is a lack of a unified national constituency. Very few interests are organized at the national level to support rural development, but there are some impressively organized groups that work as obstacles. Among these are the major farm and agricultural com-

---

Box 3.3.

## Minimalist Rural Development Policies

Many federal programs directly benefit rural areas. But we have no comprehensive rural development policy, unless the very absence of such a policy is in fact policy. This is the argument many rural policy professionals make. Terms such as *minimalist* or *triage* are often used. Triage refers to the medical emergency practice of helping those who can be saved and giving comfort to those about to die. The contested terrain is not whether these terms are accurate in describing rural development policy, but whether such a policy achieves national goals.

Advocates of minimalist policies assume that economic development, regardless of its uneven consequences for states and regions, should determine who is entitled to government benefits. Areas that are growing are more entitled than other areas to scarce development resources. Those areas left behind should not receive government economic development aid because the monies would be better spent in areas that are growing on their own. Comfort should come by way of programs that reduce the pain but do not discourage mobility. Critics of this perspective argue that this amounts to a self-fulfilling prophecy in which those places once left behind will continue to be left behind.

Advocates of a triage policy contend that people will eventually leave declining areas in search of better economic opportunities. Opponents point to the persistence of such places as pockets of poverty and, more recently, the sites of toxic and other types of waste operations. Consequently, opponents argue, there are social and environmental costs to neglecting these places.

---

modities groups, who probably correctly see monies for rural development as eventually coming from farm price-support policy. The land-grant universities' colleges of agriculture, which have a congressional mandate to assist all rural people, continue to devote few resources to rural nonfarm issues relative to their farm-based expenditures.

Alone, eliminating the myth of farm dependency will not open the way for a coherent national rural development policy to emerge. But it will remove an important stumbling block for change. It also suspends one of the pillars of farm policy, the misplaced argument that farm programs effectively provide a de facto rural development policy.

Even a move to a national rural policy would be at best but a small step forward. Given the importance of locality and regional political economy, rural development is more a local and a state responsibility. What could a national effort really do for rural communities? The federal government's role could best evolve into efforts that reduce the consequences of regional and geographically uneven economic development, particularly in the provision of education and health services.

Finally, a good case can be made that programs which facilitate the development of rural nonfarm economies may also form the basis of a good small farm policy. This point was aptly made in a Doonesbury cartoon (Figure 3.10). The dependence of small farming operations on off-farm income suggests that continued opportunities, especially an improvement in the number and quality of these opportunities, can enable these farms to continue.

*Farm policy should be taken for what it is, namely, industrial policy with some economic benefits for farmers and their industrial partners.*

The myth of rural dependence on farming is not only inconsistent with the socioeconomic realities of most rural areas; it also serves as an obstacle to the emergence of nonfarm rural development programs. Continuing the myth affords legitimacy to commodity programs that have little impact on most rural people. It also hides or plays down the existing transformation of the rural economy and society. The trailer park and the manufacturing plant are more characteristic of rural people's lives and livelihood than are family farms. To the extent that a large proportion of farm families rely on off-farm income, we are now missing the best opportunity for helping that plurality by facilitating the development of more viable nonfarm rural economies. What we urge is a simple understanding: farm policy should be taken for what it is, namely, industrial policy with some economic benefits for farmers and their industrial partners. The remaining chapters continue to make this point.

Figure 3.10.
*Doonesbury*©,
by G. B. Trudeau.

# 4 Never Base Decisions on the "Average" Family Farm

Jefferson and Emerson wrote of farming in the most misleading of ways, as if patterns of small land holdings and diversified family farms were universal for all of American agriculture. This was not true in their time, and it is even less true in ours. Discussing the average-sized family farm is very much like discussing the average-sized U.S. family. When calculated to draw comparisons, averaging either one has flaws that are all too apparent. No one has 1.8 children living at home, as the 1990 Census reports. Consequently, that particular average reveals little. Although some farmers actually did produce on the national average of 467 acres reported in 1991 or earned the average farm income of $28,887 reported in 1990, the figures are no less meaningless. They fail to describe the wide diversity of U.S. farms, and they underestimate the production capacity of those who depend on farming for most of their family incomes.

## The Fallacy of the Average Farm

Why are farm averages so misleading? At least four very important differences among farms help us understand why:

1. Differences in farm size
2. Differences between regions of the country
3. Differences in the use of inputs
4. Differences in characteristics of the operator

These differences are as complex as the differences among rural communities discussed in the previous chapter.

**Table 4.1.**
**Farms Vary by Size, 1990**

| Value of Sales Category | Number of Farms (1,000) | Percent of Farms | Percent of All Cash Receipts | Percent of All Government Payments |
|---|---|---|---|---|
| Less than $40,000 | 1,513 | 70.7 | 8.8 | 8.2 |
| $40,000 - $99,999 | 306 | 14.3 | 12.4 | 18.9 |
| $100,000 - $499,999 | 278 | 13.0 | 34.1 | 57.6 |
| $500,000 and over | 43 | 2.0 | 44.8 | 15.3 |

Source: USDA.

First, farms vary by size. The Census of Agriculture defines a farm as "any place from which $1,000 or more of agricultural products were produced and sold or normally would have been sold during the census year." Under this generous definition, we have more than two million farmers in the United States. Note that the requirement is $1,000 in sales—not $1,000 of income. Since net income is generally one third of gross farm sales, it takes roughly $40,000 in sales to generate an income that begins to be adequate to support a farm family. Nearly three quarters (70.7 percent) of the farms in the United States had sales of less than $40,000 in 1990 (Table 4.1). Note that even though there are more than two million farms, 321,000 produce 80 percent of all farm output.

Before we become too concerned about those with sales of less than $40,000, keep in mind that the vast majority of these farmers have off-farm income sources (including social security for many who are semiretired). In fact, nearly half of all farms had sales of less than $10,000—so small that many in agriculture would be hard pressed to call them farms.

Second, farms vary by region of the country. Table 4.2 demonstrates this point; for example, the average Vermont farm was 214 acres compared with 2,453 acres for the average Montana ranch in 1989. Clearly, comparing Vermont with Iowa farms and either of these with the size of a Montana ranch is silly. Doing so disregards the different commodities produced in each state and the disparities of scale in agriculture needed to do so.

Third, farms also vary by what agricultural specialists call input needs, or what farmers use to grow something successfully on a given plot of ground. As Table 4.3 shows, comparing gross commodity sales in Texas with those in Nebraska is not appropriate. In Texas, costs for inputs are high because of less fertile soils, and in Nebraska, richer soils allow farmers higher yields. In short, the costs of production vary widely across states, and this has interesting implications for public policy decisions made to assist farmers (see Chapter 6).

Table 4.2.
**Farms Vary by Region of the Country, 1989**

| State | Size of Farm in Acres | Average Net Farm Income | Government Payments per Farm |
|---|---|---|---|
| California | 373 | $70,983 | $4,430 |
| Georgia | 263 | 25,308 | 3,606 |
| Illinois | 331 | 25,503 | 8,441 |
| Iowa | 319 | 22,984 | 9,345 |
| Kentucky | 148 | 11,772 | 1,246 |
| Massachusetts | 99 | 29,536 | 565 |
| Montana | 2,453 | 18,623 | 11,716 |
| Nebraska | 826 | 36,709 | 9,514 |
| Vermont | 214 | 15,943 | 1,014 |

Source: USDA.

Table 4.3.
**Farms Vary by Input Use**

| State | Cost of Producing Corn 1988 (Dollar per Bushel, Including Land Charges) | 1989 Productivity Index (1977 = 100) |
|---|---|---|
| California | $4.17 | 143 |
| Georgia | 3.42 | 152 |
| Missouri | 4.02 | 131 |
| Nebraska | 3.66 | 125 |
| Texas | 5.49 | 103 |
| Vermont | 4.74 | 116 |

Source: USDA.

Fourth, farms vary according to the characteristics of those who operate them. Dramatic differences are apparent in Table 4.4. Comparing the farm income of a part-time producer who holds yet another job with that of a full-time producer explains little about how much farm income contributes to an individual family's quality of life. Sales of $10,000 may make a real difference in lifestyle gains to a part-time farm family, while sales of $200,000 may not provide enough cash flow for living expenses for a full-time farm family burdened by debt.

Despite the inherent inaccuracies, farm acreage and income averages are still used. Because of this persistence, both data users and the

Table 4.4.
**Farms Vary According to Operator Characteristics, Average for 1985 to 1989**

| Sales Category | Direct Government Payments per Farm | Net Cash Farm Income per Family | Off-farm Income per Family[a] |
|---|---|---|---|
| Less than $10,000 | $ 393 | $ – 1,017 | $ 29,966 |
| $10,000 - $19,999 | 1,898 | 3,185 | 25,842 |
| $20,000 - $39,999 | 4,774 | 10,056 | 22,961 |
| $40,000 - $99,999 | 9,769 | 25,929 | 18,016 |
| $100,000 - $249,999 | 18,556 | 64,631 | 15,925 |
| $250,000 - $499,999 | 28,467 | 134,367 | 21,658 |
| $500,000 and over | 36,641 | 572,219 | 27,925 |
| All farms | 5,568 | 23,672 | 25,355 |

[a] Off-farm income is reported per farm family only, not for the farm operation. As farm size increases, there are generally more multiple-family farm operations.

**Source:** USDA.

public at large have a distorted image of the family farm. Family farmers are viewed as a nationally homogeneous class of Americans, when in fact they are not. They are also perceived as much more economically disadvantaged than they actually are. USDA data show that small farms, despite limited acreage and farm income, have relatively high off-farm income. Even farms with sales of less than $10,000 averaged $29,966 per year in off-farm income from 1985 to 1989.

The common approach to reporting farm income is to compare the average income from farming with the average income for nonfarmers. This bit of evidence is supposed to demonstrate that farmers are at an economic disadvantage. If the off-farm income earned by farmers is not included in these presentations, they can be alarming. Operators of small farms will have very low farm income, as seen in Table 4.4. But it is usually supplemented with off-farm earnings. For those who wish to compare the income of farmers and nonfarmers, the more reasonable approach is to consider the household income of farmers by including what they earn off-farm. If farm income for operators of the largest commercial farms is compared with nonfarm income, a very different picture emerges.

*The many uses of averages for U.S. farms lead to inaccurate perceptions of U.S. farms.*

The many uses of averages for U.S. farms lead to inaccurate perceptions of U.S. farms. Special interest groups have marshaled these misperceptions to protect their public policy agendas. Let us examine the numbers in Table 4.4 more closely. The average net cash farm income is the number that is most commonly used to communicate "the farm problem." According to the numbers in this table, that value averaged $23,672 over the 1985-89 period. Just for a moment, consider what would happen if the definition of a farm changed and only those who had $10,000 or more in sales were considered farmers (a reasonable new definition for most purposes). Under these conditions the average five-year net cash farm income was $46,734. This estimate communicates a very different message about farm well-being.

The distorting effect of commonly used averages creates the widely held impression of a weak family farm structure, one that is easily threatened by corporate farming. As a public policy response, farm programs are seen as a necessary shield for the average farmer whose life and work are threatened by corporate takeover. The rhetoric of those who use averages to defend family farming makes it appear that corporate raiders lurk just around the corner, waiting to steal the last real farm in North America. As we explained in Chapter 2, however, they have not yet arrived, nor have they affected farm structure conditions.

## Using the Math of the Average Farm

In agriculture, policy debates historically have been general. This tendency has been linked to agrarianism. In Chapter 2 we reviewed Jefferson's argument that a strong democratic society is based on an independent, landowning farm community. The extension of that belief has long been that public policy should aim to thwart the creation of an American aristocracy that would concentrate economic power. Jefferson argued that it was not the owners of large estates (like himself) who held the hope for American democracy, but average small holders. Those who followed his prescription for democracy perpetuated the myth of the average American farmer.

As observers of U.S. farm policy have noted for decades, decision makers in agriculture have been beaten about the head by the club of agrarianism. Every time any issue is perceived as a loss for farmers, Jefferson's yeoman reappears in relatively modern dress as an endangered defender of democracy. Who can vote against *him*?

From Maine to California, a particular breed of individual has been held out to public officials and citizens as an endangered subspecies of humanity. Mr. Everyman Farmer has been embodied nationally in an image constructed of bill cap, dirty boots, blue jeans, big belt buckle, dark facial tan under snow-white forehead, calloused hands, friendly smile, and eyes made desperate by a hostile economy. Average statistics serve Mr. and Mrs. Farmer and their advocates well, because few U.S. citizens ever learn of the actual size of Iowa farms, the prevalence of part-timers in agricultural production, or the decreasing importance that farming plays in large parts of the country.

In a society where farm roots are but distantly remembered in family legend and where few people actually come in contact with farmers, stereotyped homogeneity is a prime generator of public interest and thus public policy support. Even in Congress, virtually no one studies the intricacies of farm policy with its arcane language of baselines and target prices. But concerned observers have seen media portrayals of the farm image used for advertising, they do understand the importance of eating regularly, and they can identify with what appears to be low incomes even though the concept of an acre proves a mystifying measure of property. Thus policymakers and the public are bedeviled, albeit vaguely, by the fear that something quite bad will happen if we lose even one more farm family.

After years of rhetoric about protecting the family farm and of careful construction of the heroic farmer image, informed public policy is left hamstrung by averages. Consumers of information are misled both by understatements of the strength of the *biggest contributors* in the farm sector and by overstatements about which farmers are *signifi-*

*cant contributors* to the economy. For many, the very large California fruit producer is assigned the same status as the truly poor Kentucky tobacco tenant farmer.

The mood of support for the average family farmer has provided strategic advantages to many economic interests looking out for their own: commodity organizations that want to protect crop programs for farms of all size, agribusinesses that relish the availability of large supplies of inexpensive farm-grown raw materials, international grain merchants, and environmental organizations that have linked the success of conservation programs to programs that support farm commodities. There are many others. All have become stakeholders in the law of averaging and the symbolic defense of the stereotyped Mr. and Mrs. Everyman Farmer and their family.

## Why There Is No Average Farm

In 1991 there were just under 2,105,000 U.S. farms, well below the 6 million plus farms of the 1930s (even though the 1930s numbers were slightly inflated because they included shareholders). Averages do show one important thing in looking at these losses. While the average farm of the late 1930s was quite different from the average farm of today, society appears unharmed in the amount of food (Table 4.5). Exports of farm products are also up, not down. So the past, at least, has crippled neither U.S. agriculture nor the food-dependent consumer. Understanding why is important.

Table 4.5.
**Changes in Farm Numbers, Food Consumption, and Agricultural Exports Between 1940 and 1988**

| Characteristic | 1940 | 1960 | 1988 |
|---|---|---|---|
| Number of farms (thousands) | 6,097 | 3,704 | 2,159 |
| Index of total food consumption (82–84 = 100) | 87.4 | 93.0 | 105.7 |
| Real value of agricultural exports (million 82–84$) | $2,959 | $16,216 | $32,206 |

Source: USDA.

### Not a New Crisis

The increase in yields and the expanding capacity to grow came about because of changes in equipment, innovations in plant and animal development, and improved farm management practices. Hybrid seeds, four-wheel-drive tractors, veterinary services, cost-accounting practices, and many other innovations all helped revolutionize American agriculture. These were sweeping changes. Many farmers adapted, used technological innovation, improved their education and skills, and bought or leased additional property. Those who did not adapt to the new ways either got out of farming or became those part-time operators as discussed in Chapter 3.

This progression or, as Willard Cochrane would have said, "response to the technological treadmill" can be seen in the rate of farm loss nationally (for more on Cochrane see Box 5.1 in the next chapter). Despite popular perceptions, in part gained from movies about the farm crisis, the *big changes* were not made in the 1980s. They came much earlier. As can be seen in Table 4.6, both the absolute number of farms lost per year and the percentage of all farms lost per year are in decline. Losses peaked between 1950 and 1960, the period when farms mechanized quickly following years of restricted agricultural investment during World War II. While an average of 168,500 farms were lost per year in the 1950s, losses were 29,600 annually in the 1980s. And while 12 percent were lost in the 1980s from among those that started the decade, nearly 30 percent were lost in the 1950s.

### Who Met the Treadmill's Challenge?

In the United States, 15 percent of farms produced almost 80 percent of farm commodities sold in 1990. These highly productive farms fall in the $100,000 annual sales categories or above. Only about one of every fifty farms, however, is extraordinarily large, with gross sales of $500,000 or more. USDA reports that two thirds of farms with $100,000 or more in annual sales sell less than $250,000 worth of produce. What exists, in brief, is a category of highly productive farmers who also provide good but not great incomes for their families (see also Chapter 3). They have successfully challenged technology's treadmill of necessary innovation in agriculture.

Smaller farms that produce less have not always successfully used technology. Producers in the next lowest category, $40,000 to $99,999 in gross sales, average a *net farm* income of about $26,000 per year, or only 40 percent of that earned by those in the $100,000 to $249,999 category. The typical producer in the $40,000 to $99,999 category also earns 40 percent of *total cash* income from off-farm sources. Quite

Table 4.6.
**Number of Farms Lost per Year in the United States, 1945 to 1990**

| Year | Number Lost per Year | Percent Lost in Period |
|------|---------------------|------------------------|
| 1945-1950 | 63,800 | 5.30 |
| 1950-1960 | 168,500 | 29.80 |
| 1960-1970 | 101,400 | 25.60 |
| 1970-1980 | 51,600 | 17.50 |
| 1980-1990 | 29,600 | 12.10 |

**Source:** USDA.

clearly, the level of agricultural sales in this category does not provide enough family income.

## But Who Benefits from Government Programs?

The implication of the above seems simple. And indeed any conclusion drawn from that information would be *oversimplified* because it ignores several variables: higher debt-to-asset ratios for operators of larger farms, lower quality of life expectations for operators of small farms, or other factors that are basically extraneous to the *necessity* for government support. But oversimplification aside, two points seem clear. First, on average, for farms with sales above $100,000, farm incomes are enough to argue against government price supports. Second, in the aggregate, government price supports are unnecessary to sustain adequate U.S. food production, because operators of large farms generally do not need—although they may want—farm programs.

If the above seems harsh, other data should be kept in mind. First, 73 percent of farm program benefits go to only 15 percent of the largest farms. Second, according to USDA estimates, only 32 percent of farms in all sales categories raise program crops and depend on direct government support. An economically viable livestock industry receives no direct price-support assistance. Government provides other, less costly programs to care for problems encountered by those producers. While the need for animal inspection may be debated, the lower cost of these programs as compared to commodity price policy is clear. In 1989, direct producer payments totaled $10.6 billion. This compares to less than $0.4 billion for plant and animal disease control.

## What the Myth Also Hides

As James Shaffer found, operators of large farms already high in net cash farm income get the bulk of the billions paid out in direct payments. This four-point system of deficiency payments, loan rates, conservation reserve, and disaster payments is not a safety net for the small farm family or anyone plagued by fear of its demise. But popular belief in a farm safety net, as misleading as it is, reflects a vestige of truth more real than other strands woven into the myth of averaging. Payments behind the safety net do distribute usable extra farm income that did help some producers survive the financial stress of the 1980s.

### What Is Not Happening in Farming?

The horrible fears of a world without the "average" U.S. farmers often revolve around who will farm after their demise. What if faceless corporations, such as banks and insurance firms, do eventually control farming? Given the resulting concentrations of production in large farms, the story goes, price fixing and consumer gouging will run rampant. Worse, the fear goes on, will occur if the corporations are foreign owned or if foreigners are allowed to own U.S. farmland. Yet foreigners owned only 1.3 percent of all the U.S. farmland in 1988.

Even though many family farmers have incorporated their own farms for tax and other business purposes such as personal liability, less than one third of one percent of all U.S. farms are non-family-owned corporations. Only 6 percent of farms in the very largest sales category are held by such corporations. Indeed, farm owners personally operate 89 percent of these largest farms. Tenants are slightly more likely to operate farms than are owners in each of the three sales categories between $10,000 and $250,000. So if small farms disappear, we have every reason to believe that independent farm operators would still provide the bulk of farm production.

*Less than one third of one percent of all U.S. farms are non-family-owned corporations.*

### Noncommodity Program Interest in Agriculture

Equally damaging to sound public policymaking is the popular view that the heroic Everyman Farmer dominates food production. In reality, relatively few farmer hands are actually the ones responsible for putting food directly on grocery shelves or breakfast tables. The food and fiber industry, or agriculture at large, is responsible for approximately 15 percent of the U.S. Gross National Product. More than 85 percent of that amount owes to contributions by nonfarm businesses.

In public policy, however, the vast amount of money and policymaking attention that goes to farm problems rather than to all agricultural problems leaves the average farmer looking more like an opportunist than a hero. While some of the congressional bargains that have been cut result in some equity, equal attention has not been given to all facets of the food industry. Rather, farm-state legislators have gotten votes from urban legislators by allocating dollars in approximately equal amounts to farm payments and food stamps. And African American legislators generally have agreed to vote for commodity programs so that southern legislators will vote for civil rights bills.

But what gets left out as such bargains are transacted? Congress is left open to charges of massive neglect due to budgetary constraints and even the lack of time to consider fully all problems in the food and fiber system. Policy made on behalf of the average farmer generally neglects the broader set of individual firms and institutions that have provided farmers with the means to grow, convert their raw materials to products, and find uses for these products. A few examples illustrate the point. One of every three farm workers today is not an operator, but rather a hired employee. These employees earned an average of $5.36 per hour in 1988. Also, land-grant universities, which have provided both technology and educational assistance to farmers, receive a shrinking share of the federal agricultural budget and increasingly develop new technologies with private sector support. Whether such technologies serve the public's interest or focus on the most important social needs is often unquestioned.

## Should Government Forget the Ideal of the Family Farm?

Because stereotyping farmers has worked against producing an informed electorate and developing acceptable public policies, should the ideal of family farming be rejected? By no means. As noted earlier, most farms are operated by farm families, often in partnership with other families. They do work hard, long hours; few of them are truly wealthy; and few of them have retirement funds for their old age. The misleading use of the "average" farmer and the image that goes with it does not negate the importance of family farms. But not, we argue, for the reasons used to justify past support.

Our complaint is twofold. First, present policies, which have become outdated since they were established in the Depression Era, should not be perpetuated only because averaging makes farm incomes appear lower than they actually are. Second, society's focus on farmers as sole sustainers of U.S. agriculture diverts resources to them at the expense of other problems in the food and fiber system.

We argue first that the American farm be seen for its diversity and variety. Farming is not the same across the United States. And producers in different regions face as many unique problems as they do ones common to national circumstances.

We also argue that if family farms gain future financial support it should be because of the value that society places on that type of farm structure. If we are to support farmers, let us use a social argument, not thinly veiled economic arguments. Smaller farms where families must rely on second jobs to continue production may be worth preserving because they keep rural America alive and lend a purpose to what goes on there. For most rural areas, however, the best support for operators of small farms will come through rural development to improve off-farm job opportunities rather than through government involvement to improve farm incomes (see also Chapter 3).

*If we are to support farmers, let us use a social argument, not thinly veiled economic arguments.*

Operators of larger farms, if they gain continued price supports, should receive these benefits either (1) because farming is a risky enterprise made vulnerable by weather or (2) because maintaining them can hedge against a time when world food needs and wants may expand considerably. In the latter case, government should be paying to build up a strategic reserve. In other words, government and society may simply want a large reserve of independent farms of all sizes and shapes to provide a flexible and adaptable farm sector. Any such goal, if it exists, should be more carefully spelled out to those who pay the bill. Only if this preference is specifically stated can the United States justify farm benefits used as, for example, an economic development package that keeps 60,000 small farm businesses alive in Nebraska.

Six characteristics work together to make the farm sector economically unique (see Box 4.1). But beyond these characteristics, U.S. agriculture is not corporate in structure, national in its production problems, or intimately linked by the diverse commodities produced. As a consequence, any national policy that supports family farming directly must have a strong regional, if not state, orientation. It must recognize that income needs and market influence vary by farm size and type. Policymakers must be aware that budgetary trade-offs exist between supports for farmers and funds for addressing other issues in the food and fiber system. And it must hold that family farms have social (perhaps agrarian?) as well as economic and consumer value.

More than anything, the myth of the average family farm persists for one single reason: to justify the costs of protecting farmers. Those who fear that lack of honesty may well be correct. For both public officials and the citizenry, it may not matter whether the portion of their disposable family income that goes to food ultimately ends up in the pockets of a family farmer in Nebraska, a corporation in California, or some equally unknown producer in Brazil or France.

Box 4.1.

# There Are No Average Farms, but There Are Common Characteristics of Agriculture

Six characteristics, none of them individually unique to production agriculture, combine to make the agricultural sector economically unique. Two are characteristics of the demand for agricultural products; four influence the supply of agricultural products.

## 1. Growth Is Slow

The demand for food grows as the population grows and as per capita income increases. In the United States, the population is growing at only about one percent per year. Median family income is already quite high at about $34,200, and only about 12 percent of disposable income is spent on food. As already high per capita income increases, a decreasing part of this increase is spent on food. Quite simply, there are limits on how much food people can eat. Estimates indicate that for a one percent increase in per capita income, consumers would increase food expenditures by less than two tenths of one percent. Thus, by combining population growth and increasing per capita income, we see that demand for agricultural products (as opposed to food) in the United States is increasing only slightly more than one percent per year.

## 2. Demand Changes Are Sluggish to Price

The quantity of agricultural products demanded, taken as a whole, is relatively unresponsive to changes in price. That is, if the price increases, the quantity demanded will fall less than proportionally. If the price decreases, the quantity demanded will increase, but again less than proportionally. This response is important for two reasons. First, the five decades of price-support programs in agriculture have depended on this characteristic. With a price support above levels that would clear the market or use all goods, consumers decrease their purchases less than proportionally to the price increase. As a result, total revenue to the farmer from the market increases. Thus the government, as the purchaser of last resort in the market, can support the price with relatively modest purchases into government stocks. But the growing influence of international markets, as discussed in Chapter 7, has reduced the influence of this demand characteristic. Second, relatively small changes in supply will cause relatively large fluctuations in price. A bumper crop will cause prices to plummet. And a short crop due to adverse weather or pest infestation will cause more than proportionally strong prices.

## 3. Nature May Rebel

This characteristic—nature—brings us to those characteristics that affect the supply of agricultural products. Climate dictates the geographic limits for production of specific crops and, to a lesser extent, livestock. Weather can cause

➤

significant changes in production levels from one year to the next. Pest infestations can reduce yields significantly. Because of nature's vagaries, agricultural prices and incomes often become quite volatile.

### 4. Farms Are to Agriculture as Atoms Are to the Body

Because we have a great many independent farms—that is, farming has an atomistic structure—producers are "price takers" for their products. No single producer can affect the market price by withholding products from the market or dumping excess products into it. Thus individual farmers, for their own self-interest, produce and market as much as possible. This, in turn, may not be in the best interest of agriculture.

### 5. Technology Moves Quickly

Public sector and, increasingly, private sector investment in agricultural research and development has been highly successful in producing a steady stream of new technologies: improved seeds, fertilizers, crop and animal protection chemicals, labor-saving machinery, land-saving cultural practices, and irrigation systems. These have made agriculture ever more productive. Because of farming's atomistic structure, individual producers adopt these new technologies as quickly as possible to lower costs on each bushel, gallon, or animal produced. This is a proven way to increase individual production and marketings. But consumers are the ones who ultimately benefit most from new technologies.

### 6. Investments Are Trapped

For the most part, land in agricultural production has little value if used for other things or if idled. Specialized capital—tractors, tillage equipment, milking parlors, combines—cannot be converted easily, if at all, to nonfarm uses. Even labor, once committed, often cannot readily move out of agriculture. This is because many nonfarm jobs are distant from the farm or require skills that farmers lack. Age sometimes acts as a barrier to farmers moving out of agriculture. Thus resources—what economists call assets—once committed to agricultural production tend to remain there even when the economy flounders. This goes on despite the long-term downward trend in prices for the sector.

The interaction of these six characteristics causes behavior peculiar to agriculture. The slow growth in demand and the relative unresponsiveness of demand to price, coupled with the trend of rapidly increasing supply prompted by the development and adoption of new technology, cause real prices of agricultural products to decline over time. Contributing to the tendency for overproduction and declining real prices is the fixity of resources, particularly land, in the sector that keeps more resources in production than the market needs.

Adapted from Dale E. Hathaway, *Government and Agriculture*, 1963.

# 5 Never Confuse Production with Productivity

Farm program benefits are tied to farm production. This program design stems from a longstanding tradition of equating productivity with moral virtue. Unfortunately, the pursuit of greater production encourages more output than the market can absorb, with, occasionally, damage to the environment. The problem is that we have confused moral virtue with productivity. In this chapter, we distinguish between the moral virtue of hard work and the benefits from gains in productivity. This distinction allows us to see how productivity can be helpful to farmers and consumers in the right policy environment.

## Production as a Moral Virtue

Two subtle but easily recognized strains of American culture have reinforced the belief that a farmer's moral character can be measured in terms of total farm production. One is the virtue of hard work, discussed in Chapter 2, a virtue that encouraged more total production, given the conditions under which 18th and 19th century Americans farmed. The second is the importance of fertility and fecundity as signs of moral excellence for Calvinist sects that populated rural America. Production continued to be a cultural value for our farmers long after the explicit theological rationale faded into obscurity.

### Hard Work

The first theme—the relationship among production, moral virtue, and hard work (or industriousness)—is easily grasped. As discussed in Chapter 2, work was considered to be a moral virtue among 19th century Americans, as it was for many other cultural groups at various times and places. The rewards of work included personal satisfaction and growth, but they also included material success. This meant that work was expected to produce material abundance, and for a nation of farms that meant more food and fiber commodities. More supply was good for the farm household and the rest of society. Both for personal use and for sale on expanding markets, increased production was a clear sign of moral virtue. People who worked hard were thought to be morally good, and the visible evidence of work was found in what they produced. Production, itself a reward for hard work, was also a sign to others that the producer was hardworking and virtuous.

> *Production, itself a reward for hard work, was also a sign to others that the producer was hardworking and virtuous.*

### Fecundity

Protestants with Calvinist roots believed that God's chosen would live moral lives. They also believed that God revealed His chosen people through visible signs.[1] One sign of being among the elect was fertility or, better, fecundity. Fertility and fecundity imply a capacity to produce abundantly, particularly with respect to natural reproductive processes. This quite simply meant that the elect would have many children. By extension, the capacity for producing many progeny would also be seen in a farmer's domestic stock and the plants under cultivation. The theme of fecundity differs quite sharply from that of hard work, since fecundity is neither a reward for earthly activity nor a sign of any virtue that can be won by human effort. It is a sign of God's grace, a sign that one is a certain kind of person, one of God's chosen.

The religious doctrine of fecundity of the elect is the root of an attitude that continued to permeate the self-image of rural Americans, even after Calvinist doctrines became diluted and mixed with other strains of Protestant religious belief. Those who produce children, animals, or crops take a certain comfort and justification in their success. Their fecundity is evidence that God's grace has smiled upon them. It is unthinkable that one would not do what was required to bring forth God's bounty. To fail to produce the biggest and the most would be to refuse God's grace—and that would be an act of sinful pride, an act of turning one's back on God. The practice of producing the biggest and the most became a foundational part of the farmer's identity. It was so fundamental that it could not be questioned.

"This is God's country" signs dot the Midwest, continuing to acknowledge God's grace in securing the productivity of His people. The productivity of the land itself is not due to human effort, and the bounty produced from it is not earned. But because God has given His people productive land, they are entitled to the food and fiber commodities that can be produced from it. Indeed they would be ungrateful if they did not fully accept the gifts of grace.

*Moral Virtue*

Although the virtue of industry and the doctrine of the elect provide rather different accounts of why more production is a good thing, they share several important characteristics. First, farmers who believed in either value had reason to raise both larger quantities and bigger specimens. There was moral value in both higher yields and producing the largest hog, pumpkin, or whatever. Second, the larger production and bigger specimens were not valued for themselves. Rather, the farmer's ability to produce them was taken as evidence of some underlying, perhaps hidden, virtue. Because the values sought in pursuit of more production are implicit, they were easily obscured and relegated to the class of myths. Third, the desire to produce more and bigger had absolutely nothing to do with a farmer's ability to sell or consume the commodities produced. As evidence of a person's moral virtue, these commodities had a value unrelated to their food value or to the price they would bring. This final point places the moral value of greater production into sharp contrast with the concept of productivity developed in production economics.

## Production Versus Productivity

*Productivity* is the *rate* at which inputs such as labor and raw materials used in the production process are transformed into outputs such as finished goods or, in the case of farming, food and fiber commodities. The main idea behind the concept of productivity can be seen in the simplified equation:

Inputs (labor, land, capital, materials) × productivity =
production (all outputs)

Production can be increased by increasing either the inputs or the productivity. If a farmer increases the amount of land being farmed or works longer days, production can be increased even when economic productivity remains constant. Productivity is increased normally by changes in technology. A new seed variety may allow the farmer to get more production from the same amount of land, labor, fertilizer, and other inputs. An increase in productivity can also be used to reduce the inputs so that land, labor, energy, or other resources used in farming can be used to produce something else.

It is worth noting how the moral virtue of hard work differs from productivity. Working harder is simply a way of increasing labor inputs, either by working longer hours or by increasing the intensity of work. In technical terms, the increased production that comes from hard work is due to an increase in labor inputs, not to an increase in productivity. The laborer who simply works longer hours produces more because inputs are greater, not because he or she has greater productivity. The moral value of an individual's willingness to put more work into the production process is not the same as an increase in productivity.

In fact, increases in productivity have almost certainly tended to undermine the work ethic. If a farmer gets more production from less work (perhaps because of new seeds or machines), production can hardly be regarded as reward for hard work or as a sign of moral virtue. The link between the virtue of work and increased production is not broken altogether, however, because more skills may be needed to manage a new technology. Since the advantages of new technology go to those with more education and the resources to invest in the technology, economic power becomes more important as productivity increases. The hardworking poor may be left out of productivity innovations. If those left behind believe that production and wealth should be the reward for work, as opposed to capital, they will probably take a jaundiced view of productivity gains.

## Why Increasing Productivity Is the Solution, Not the Problem

The productivity of U.S. agriculture has clearly grown over the last forty years (Figure 5.1). More output is obtained from fewer inputs

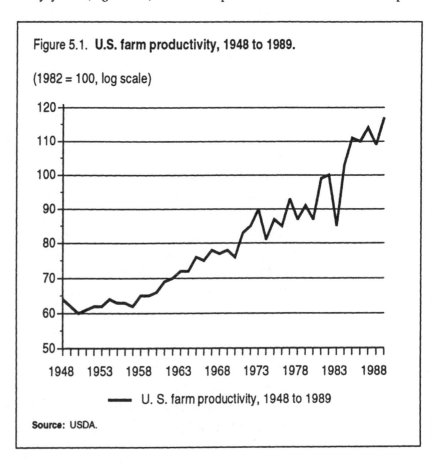

Figure 5.1. **U.S. farm productivity, 1948 to 1989.**

(1982 = 100, log scale)

—— U. S. farm productivity, 1948 to 1989

**Source:** USDA.

because of improved technology such as mechanization, better seed varieties, and use of pesticides. These changes allow more food and fiber to be produced with less labor and land.

This increased productivity has been blamed for a wide variety of problems in agriculture, from the alleged disappearance of the family farm to chemical pollution of the environment. Many of the undesirable events often associated with increased farm productivity have nothing to do with productivity as such. To understand this, consider four common criticisms of productivity.

### Disappearance of the Family Farm

New technologies increase output and lower costs. This increases supply, which reduces price and eliminates the initial profits gained from adopting a new technology. Because less labor was needed with new technologies, fewer farmers were needed to produce the same amount of food. Farmers who reaped the benefits of adopting a technology early expanded their operations as others left farming. Thus technology, through the treadmill, was blamed for creating a smaller number of larger farms over time (see Boxes 5.1 and 5.2).

Technology made it possible for fewer farmers to produce food in the United States. Is this undesirable? Would this be a better country if we all grew our own food? Romantics might answer yes, but most people understand that fewer farmers means that more people can work in industry and services. The total economic output of the U.S. economy grows whenever we can produce more goods with fewer inputs. When productivity increases in agriculture, then agricultural inputs can be used to produce other goods.

Agriculture's income problems do not arise from the treadmill of new technologies, but rather from the fact that demand for food has limited growth potential. When more food can be produced with less, the result is often sharply falling food prices. Since domestic demand for food is limited in a high income country like the United States, increased productivity typically means that fewer inputs (including farmers) are needed to produce food. The ability of U.S. agriculture to export food to other countries has slowed the process of reducing the resources used in agricultural production (see Chapter 7), but this process is still an inexorable part of economic growth.

### Chronic Overproduction

If the land, labor, and material resources being used remain constant, then increases in productivity lead to increases in total production. When total production exceeds demand, we have overproduction. But production can be brought down by reducing inputs.

## Cochrane's Agricultural Treadmill

"The aggressive, innovative farmer is on a treadmill with regard to the adoption of new and improved technologies on his farm. As he rushes to adopt a new and improved technology when it first becomes available, he at first reaps a gain. But, as others after him run to adopt the technology, the treadmill speeds up and grinds out an increased supply of the product. The increased supply of the product drives the price of the product down to where the early adopter and all his fellow adopters are back in a no-profit situation. Farm technological advance in a free market situation forces the participants to run on a treadmill.

"For the laggards who never got on the treadmill, the consequences of farm technological advance are more devastating. Farm technological advance either forces them into a situation of economic loss or further widens their existing losses. In the long run, such sustained losses must force them into bankruptcy and out of the business of farming.

"As these laggard farmers have been forced out of business by the process of farm technological advance, their productive assets have typically been acquired by the better, more aggressive farmers—by the "early-bird" farmers who prospered from the temporary gains of the early adoption of the new and improved production technologies.

"Not only did the process of farm technological advance force the participants in the process onto a treadmill, but it created a condition in which the strong and aggressive farmers gobbled up the weak and inefficient. The process of farm technological advance has contributed importantly to the redistribution of productive assets in American agriculture in which commercial production has been, and continues to be, concentrated on the larger farms."

Quoted from Willard W. Cochrane, *The Development of American Agriculture*, pp. 389–390. Reprinted by permission of the University of Minnesota Press. All rights reserved.

The process of reducing resources used in agriculture can be painful and socially disruptive, so farm groups have sought government support for farm prices and incomes. The U.S. government purchases some commodities at a fixed support price, and then must dispose of this "surplus." Or producers of some commodities are given income-support payments linked to production. By providing continued incentives for production, government programs have caused more production than the market can absorb. But there is no reason why

Box 5.2.

## Are Larger Farms More Productive?

Do larger farms benefit from economies of size, as do other large-scale businesses elsewhere in the economy? For as long as this question has been asked, subsequent research consistently finds that the answer is no. A recent study by Mary Ahearn and colleagues reports that production efficiency, as measured by lower costs per bushel, levels out very quickly as farm size increases. This suggests that it is not the *rate* of profit but the *volume* of profit, or capacity to generate income, that drives farm size. Large farms provide enough profit for a household income; small farms must seek off-farm sources of income. Thus bigger is not equal to better (more efficient) for most types of farming operations, but bigger farms do sustain a household living.

farm income-support or adjustment assistance cannot be designed to avoid encouraging surplus.

### High Cost of Government Programs

If income and price supports are pegged to total production, farmers will have an incentive to produce more than the market can absorb. Increases in productivity will help them do this, but so will increasing the amount of production inputs. An increase in either productivity or inputs will be followed by an increase in government program costs as long as policies that tie payments to total production remain in place. But whether these policies do remain in place is a matter of political choice.

### Pollution and Soil Loss

Technologies that increase yields may also increase the production of unwanted outputs such as chemical pollution of groundwater or soil erosion. The catch is that the equation used earlier in this chapter obscures some of the difficulties in measuring productivity. Because we tend to think that only salable commodities are the product of farming, we have a natural tendency to exclude outputs like pollution when we calculate production totals. A more sophisticated production equation might look like this:

The value of inputs (land, labor, capital, materials) × productivity = the net value of outputs (value of salable commodities minus value of pollution)

Farmers do not pay the costs of all the pollution created by agricultural production (see Chapter 8). Some costs, such as those associated with soil erosion, come out of a farmer's future earning potential, and so most farmers will avoid depleting their own natural resources if they are aware of it. But other costs, such as pollution of groundwater or surface water with chemicals, may be less obvious to individual farmers and have consequences for other members of society.

Farmers may benefit from a new technology that increases commodity production as long as they need not pay for the costs of any pollution generated by the technology. In this case, policy *should* make the net value of production obvious to farmers. This could be done through regulation of polluting practices or through taxing polluting inputs. If a new technology is not profitable for society as a whole, then the role of policy is to make it unprofitable for individual farmers.

### Reaping the Benefits of Higher Productivity

Managers attempt to increase productivity in every industry. Increasing productivity is good for the firm because it helps the firm be more competitive. A firm with higher productivity will have lower costs, calculated on each unit of output. Such a firm can benefit either by capturing higher profits or by being able to make an adequate return on products sold at a lower price. In a competitive economy, it is the latter event that is most likely to occur. Why? Because market forces will encourage every producer to achieve higher productivity, and lower prices will be the eventual result.

This result means that society also benefits from higher productivity, because many of the savings in production costs are passed on to us as consumers. This lesson from economics is so basic that it has not only been widely accepted; it has also become trivialized. The 20th century has taught us that a simple-minded linking of productivity to social welfare is *too* simple. Producers have incentives to exploit inputs such as soil and clean water. Yet it is a mistake to treat renewable resources as depreciable goods, like machinery or buildings. Thus policy should discourage unwanted products like pollution. It should not encourage food production that no one will buy.

## Productivity and Public Policy

The themes of reward for hard work and signs of God's grace reinforced an uncritical acceptance of the assumption that producing more was always in the public interest. Few farmers or policymakers would spontaneously offer these themes as a justification for tying production to program benefits today, but moral and religious ideals associated with industriousness and fecundity have persisted. Farmers are seen as deserving of government program assistance precisely because they work hard to produce a necessary product. Merit therefore becomes an issue in price-support debates, even when quite obvious surplus production results.

While we would hope that public policy would not discourage people from pursuing the virtue of moral productivity through hard work, total production has become a very poor indicator of this virtue. Some people can produce a lot without working particularly hard or well. The material rewards for this production are captured at the expense of higher prices, or taxes, for everyone. Undoubtedly, many farmers do work hard to produce their animals and crops. But at the same time, it is hardly plausible to claim that we encourage the virtue of hard work among citizens by encouraging production of more food and fiber.

*Blind faith in production has produced a production bias among rural Americans.*

Blind faith in production has produced a production bias among rural Americans. This bias is justly criticized by those who note the social and fiscal irresponsibility of farm programs that encourage commodity production far beyond economic demand. Since it has become clear that excess production entails additional costs in soil, water, and pollution, the uncritical acceptance of production among farmers has cost them many friends. But government policies do not need to promote overproduction. And they could be designed to make the benefits of environmentally desirable production practices more evident to farmers.

## Getting the Productivity Assumptions Straight

Our faith in high production has invited confusion between moral virtue and productivity. As a result, critics who would justly criticize overproduction sometimes attack productivity. Productivity is not to be blamed for the ills of overproduction. Increases in productivity, usually the result of new technologies, ought to allow valuable resources to be used for other things. Productivity gains should reduce the costs of food to consumers without necessarily reducing a farmer's income. A solid understanding of productivity shows us that when increases are achieved at the expense of future production, as when soil is depleted or water polluted, we have not achieved an increase in productivity at all.

New technology is not always the culprit when agricultural production results in pollution or soil loss. In fact, new technologies could increase true productivity by reducing pollution. The problem is that agricultural research policy tends not to look beyond commodity production as a goal. The land-grant universities' traditional mandate is "to make two blades of grass grow where only one grew before." This goal may be outdated in a high income society that has little additional demand for food and where environmental quality is increasingly valuable. Government-funded agricultural research should shift direction towards research focused on production of higher quality consumer goods, including environmental amenities and more nutritious foods.

The belief that production of food and fiber is a natural bounty that comes to us through the grace of God must undergo a reality check. There is no reason to question the religious underpinnings of faith in God's grace. But should this religious ideal be interpreted to mean that farmers should expend soil, water, and human resources to produce food that is not needed? Faith in God's grace may have made rural Americans confident that they should produce as much as they could. But is this faith any reason to pay for production through government programs? The environmental and social consequences of overproduction should provoke a rethinking of the policy implications of that religious ideal. Surely God's grace could not mean a license to despoil the environment, waste food and natural resources, and raid the public treasury.

## Notes

1. Calvinist doctrine held that one could not *earn* salvation. God's foreknowledge of all things meant that those who would enter the kingdom of heaven had already been chosen, and nothing one did in life could change this fact. The chosen or "elect" were saved, all others condemned. To the postmodern mind, this doctrine would appear to remove religious incentives for following the moral path. If sin has no *cost*, why choose virtue? For Reformation Protestants, however, morality sprang naturally from the breast of the elect; it did not need to be "chosen."

# 6 Never Confuse Farm Prices with Farmers' Income

The idea that farmers deserve a "fair" or "just" price for their product has been around for centuries. One source of this value judgment was tradition that fixed the degree of tribute or duty owed to a superior. The amount of payment was fixed by the social relationship of underlings to their lord. Like religious tithes, payments to superiors represented an expression of gratitude rather than a transaction. In more recent times, the notion of a fair price reflects a world view of fixed social and economic relationships. People are identified by their role as producer—doctor, farmer, baker, mechanic—and society is conceived in terms of stable role relationships. In this view, an economy becomes dysfunctional whenever it fails to reinforce the pattern of social roles.

A second source of the "fair price" myth is the labor theory of value, endorsed by Adam Smith and Karl Marx, but rejected by contemporary economic theorists. Put in the simplest terms, the labor theory of value states that the true value of any good produced is equal to the value of the labor expended to produce it. If one assumes that all work is of equal value, perhaps because all humans are equal, then the fair price for a good is equal to the value of the labor used to produce the good. Any deviation from this price can be thought of as evidence that one person has used a position of power to capture someone else's labor value.

These two ideas can be applied to support the judgment that non-farmers owe farmers a certain fixed amount for the fruits of their labor. This fair price reflects the social status of farms vis-à-vis others in the community and also the value of their sweat and toil. The links to Jefferson's agrarianism are obvious. Because food is a necessity, the agrarian ideal places a high value on food production and on the hard work required to produce food.

The modern economic paradigm, however, offers a contrast. It views prices simply as signals, or incentives, that tell producers how much to grow and consumers how much to buy. When costs or demands change, prices should also change. Although most people implicitly recognize that prices are determined by supply and demand in a modern economy, notions of a fair or just price persist.

For nearly sixty years, agricultural policies have addressed the income problems of farmers by supporting farm prices. In the face of such persistence, most observers reach what *appears* to be a very logical conclusion: higher farm prices mean higher farm incomes. This assumption is used frequently in agricultural debates to promote both price supports and supply control.

> In the very short run, higher prices do mean higher incomes. But higher prices are eventually self-defeating: income gets lower, not higher.

High commodity prices may seem to be the solution to farm income problems. In the very short run, higher prices do mean higher incomes. But *higher prices are eventually self-defeating*: income gets lower, not higher.

In this chapter, we discuss three themes necessary to understanding myths about the relationship between farm prices and farm income. First, income is more important than prices. Second, higher farm commodity prices create higher costs of production because these prices bring about higher land values. This means that increased cost of production wipes out income gains and that high farm commodity prices benefit landowners the most, not necessarily farmers. Third, high farm prices reduce demand for farm products, especially in international markets.

If high prices are self-defeating, then what is the essence of the proverbial "farm problem"? Policy reform, we will show in the following pages, requires us to recognize that the farm problem has changed as agricultural conditions have been altered. Legitimacy for public support of U.S. agriculture has been built around a now flawed definition that sees the farm problem as synonymous with low farm income and, usually, a need for higher prices. But, as Chapter 4 demonstrated, farmers are no longer the disadvantaged group they once were when low income was characteristic of farm life. This means that we need to reexamine both the high price assumption and the low income assumption.

The most important aspect of today's farm problem may well be the variability of farm income relative to that of the average citizen's. Farmers inescapably face highs and lows, often intense ones. As a consequence, those concerned about farmers should not focus on the level of farm income. Rather, they should look at the variance in farm income and the associated problem of variance in asset values. To understand income and asset variance as central to the farm problem, we need to know both how the structure of U.S. farming has changed and how U.S. macroeconomic policies affect farm asset values.

## Income Is What's Important, Not Prices Alone

Price is only one part of the income equation:

Income = (price × output) − (costs of production)

Price, output, and costs all influence income directly. Farmers, in other words, can make more money by growing more or by paying less to grow the same amount.

Technology influences income by reducing the cost of production. This factor helps constrain the choice of how much to grow. For example, if the total cost of production for one acre of corn is $220 and the corn yield is 120 bushels, then the cost per bushel is $1.83. Higher yields with the same investment will result in lower per unit costs of production. If price does not change, this means more income. Any rational farmer has strong incentives to choose new technologies. As more farmers adopt the technology, however, prices fall for everyone of course. The production increases that occur almost inevitably lead to lower prices.

More than sixty years ago, farmers tried to stop the trend toward lower prices by promoting the concept of parity as a basis for determining support prices. The dual effects of technology on profitability as well as on prices were ignored by supporters of parity, a concept that, although misconceived, has been elevated to dogmatic status by many agrarians. This rather odd concept was introduced by farm groups lobbying for government support in the 1920s. As Don Paarlberg explained, "Parity was perhaps more understandably defined by the farmer who observed: 'If a bushel of wheat would buy a pair of overalls in 1910-14, then, to be at parity, a bushel of wheat should be priced so as to buy a pair of overalls today'" (p. 25).

Parity assumes that, as wheat production multiplies per acre, each farmer needs the same multiple of pairs of overalls. But does a farmer wear out pants faster by farming with a bigger tractor? Or by adopting a new strain of wheat? The assumption that farm production should buy the same amount of goods over time ignores the role of new technologies in reducing cost. Wheat production per acre has increased threefold since the beginning of the century. It is truly hard to justify why the two additional bushels of wheat are needed to buy yet two more pairs of overalls today. If the cost of producing the bushel has dropped by two thirds, as it has since 1910, why does parity not go away?

But even if parity disappeared, other equally suspect arguments would still remain. For example, many farmers argue for higher price supports in order, as they say, "to cover cost of production." Cost of production is only somewhat better than parity as a basis for determining farm prices. For example, in the face of technology's influence, would "average" costs be used (see Chapter 4)? Not all farmers have the same cost of production. Those with lower costs would benefit the most from high support prices, while those with high costs would find price supports inadequate. Alternatively, would actual cost be used? That is, would government pay the biggest subsidy to those who use, or even waste, the most inputs? And—perhaps the most troublesome aspect of using costs of production—how would land be valued? Would it be set at the current market value? As we will show below, government actions do influence land values.

## High Farm Prices Will Be Bid into Asset Values

Farmers can buy or rent land to expand their operations. The price or rent they pay is related directly to the expected earning capacity of land. The value of land is the sum of rents that a parcel of land can earn, summed over time and discounted back to the present. Usually figured annually, rents are nothing more than a farmer's payback for investment in land. But if a farmer owns a tract of land, the value of that land is determined not only by what it earned last year, but also by what it is expected to earn this year, and the next, and so on—in other words, its earning capacity. Calculating a parcel of land's present value involves estimating what all of its future rents are worth today.

When government programs raise the returns to farming, farmers expect that the land's future earning capacity will be higher. This future earning capacity will be reflected in land rents, and these increased rents will be bid into land prices. Over time, farm income will decline because returns do not cover cost of production, including the cost of land that reflects the benefits of price and income support. Under these conditions, farmers redirect their attentions, they lobby for higher price and income supports, which in turn will be bid into asset values. Thus the operation of a competitive market, when farmers get what they want from lobbying, prevents higher prices from providing higher incomes in the long run.

Because government payments are reflected in land values, the main beneficiaries of price and income supports are landowners, not necessarily producers. Most of these benefits go to a select few. The top 5 percent of landowners own well over half of U.S. farmland; 30 percent of farmland is owned by landlords who do no farming. In short, government programs that result in higher land prices skew benefits towards those who are most well off and who are least likely to farm their own land.

## High Prices Shut Agriculture Out of Markets

High farm prices are not just self-defeating, they actually damage U.S. agriculture in the long run. High prices choke off growth in demand and encourage foreign competitors in a world agriculture where nations are interdependent. Clearly, high price supports reduce the amount that can be sold in domestic and foreign markets, even in the short run. But that is hardly the only problem. Government programs must then bear the costs of storing and disposing of any commodities priced too high to sell.

There are also negative long-run effects of high price supports and income payments tied to production. Because these government payments to farmers are bid into land values, the overvalued land reduces U.S. agriculture's overall competitiveness. The higher land values increase costs of production to levels that are above international prices. Competitive markets also mean that land prices will increase even in those areas not growing program crops. Therefore, these policies have implications for all sectors of farming, including livestock farmers. Foreign competitors end up supplying a larger share of the world market, and U.S. agriculture becomes increasingly dependent on government support. This leads, quite obviously, to more farm lobbying for government support, continuing the cycle of decline.

*Because these government payments to farmers are bid into land values, the overvalued land reduces U.S. agriculture's overall competitiveness.*

High price supports and the resulting increase in land rents were one reason why U.S. agricultural exports plummeted in the early 1980s, as will be shown in Chapter 7. With their political demands for higher prices, protesting farmers of that decade did not seem to understand that high land prices contribute to lost commodity demand. Or perhaps they did not want to understand simply because it was too painful for those who had paid high prices for farmland, expecting that future returns would be high.

## The Real Farm Problems:  Asset Value Risk and Cash Flow

Today's farmers are more likely to be harmed by asset value risk and uncertainty than by low levels of farm income. As can be seen in Figure 6.1, farm household income varies more than average household income. As agriculture has expanded in international markets

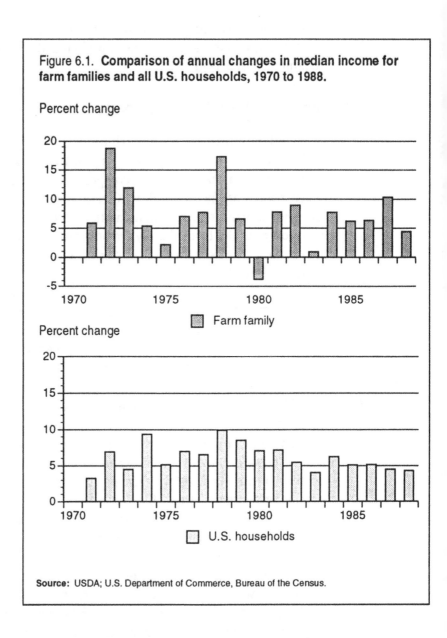

Figure 6.1. **Comparison of annual changes in median income for farm families and all U.S. households, 1970 to 1988.**

Percent change

20
15
10
5
0
-5

1970          1975          1980          1985

Farm family

Percent change

20
15
10
5
0

1970          1975          1980          1985

U.S. households

**Source:** USDA; U.S. Department of Commerce, Bureau of the Census.

and increased the use of debt, risks have intensified. For the most part, however, farmers are often less familiar with either of these risks or how asset values are related to various public policies.

### Asset Risk: The Land Market Boom and Bust

The 1970s were a prosperous time for U.S. agriculture. U.S. monetary policy was modified significantly in the early 1970s as the United States went from a gold standard to a flexible exchange rate in world markets. As a result, the U.S. dollar was devalued, which in part caused exports to expand in the early 1970s and led to higher commodity prices (see Chapter 7). Throughout the decade, inflation was relatively high. Monetary and fiscal policies focused more on economic growth than on controlling inflation. After adjustment for inflation, for reasons explained in Box 6.1, interest rates were relatively low. Inflation averaged 7.2 percent from 1972 to 1980; Federal Land Bank interest rates averaged 8.5 percent. Therefore, during this period, the real rate of interest averaged 1.3 percent, a strong incentive for farmers to borrow money. Monetary policy thus encouraged farm expansion in two ways: (1) through low real interest rates and (2) through a decline in the dollar's value that spurred agricultural exports.

The U.S. tax structure through the income tax system also provided strong incentives for farm expansion during the 1970s, when the maximum marginal tax bracket was 70 percent. Since passage of the Economic Recovery Tax Act of 1981, this rate has been between 30 and 35 percent. But when rates were high, as inflation and higher commodity prices resulted in higher nominal incomes, farmers were faced with higher marginal tax brackets.

---

Box 6.1.

## Components of Nominal Interest Rates

Nominal interest rates have three components:

- The time value of money
- An inflation component
- A risk component

Money's time value has generally been 3 to 4 percent. An inflation component is important because a lender does not want to loan money before inflation and be paid back in inflated and less valuable dollars. The risk component in part reflects the economy's faith in continued stability in monetary and fiscal policies. Bankers are unwilling to lower interest rates if they believe inflation will increase in the future. Many time adjustments are slow. When a major shock is introduced, these relationships can be abnormal for some time. For example, after the tight monetary policy of the early 1980s, it took some time for the nominal interest rates to decline, even after inflation was significantly reduced. Part of the reason was simply that the financial institutions doubted that inflation was under control, and part was because mistakes made in the early 1980s locked bankers into long-run investments based on assumptions that inflation would continue.

---

For farmers who were in a marginal tax bracket of, say, 50 percent, any tax-deductible expenses such as interest were valuable. For example, farmers who could borrow money at a nominal interest rate of 10 percent could claim this as an expense and write off 50 percent of the cost of borrowing on their taxes. Thus the nominal after-tax interest rate was really only 5 percent, which provided even more expansion signals to invest in land, livestock, or machinery.

Clearly, all signals were go for increased land prices during the 1970s. First, returns were high, and everyone expected them to remain high in the 1980s. Second, inflation was relatively high. And third, the real interest rate was low. How these factors influence land prices is explained in Box 6.2. The average annual change in real land prices

---

Box 6.2.

## Factors That Influence Land Prices

Land prices are influenced by two major components: (1) the returns, or rent-earning capacity, of the land and (2) the interest rate.

Summing future expected rents and estimating what the total value of those rents are worth today (that is, the present value) is calculated by using the interest rate, which represents the opportunity cost of money. The money tied up in land ownership could be earning interest if invested in, say, stocks or bonds. Thus the interest rate measures the cost of this lost opportunity. The relationship among the land's value, future expected earnings, and the interest rate can be seen in the following equation:

$V = R/i$

where:

$V$ = the value of the land today
$R$ = the constant rent expected in each future time period
$i$ = the interest rate

This procedure is known as "discounting back to the present" or finding an asset's "present value." For example, if a parcel of land is expected to earn a constant rent of $100 each year and the annual rate of interest is 10 percent, then that parcel's present value is $1,000 ($100/.10). Obviously, the greater (smaller) a parcel's rent-earning ability $(R)$, the greater (smaller) the present value $(V)$ of that parcel. For example, if $R$ increases to $150, then $V$ would increase to $1,500. Or if $R$ decreases to $75, then $V$ would fall to $750. Changes in the interest rate also affect land's present value. Assuming rents of $100, if $i$ increases to 15 percent, $V$ would fall to $667. This occurs because alternative investments could earn a higher rate of return than would committing the money to land. If the interest rate were to fall to 5 percent, then $V$ would increase to $2,000.

---

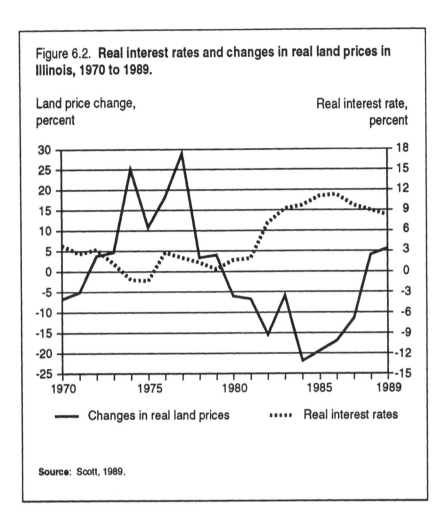

Figure 6.2. **Real interest rates and changes in real land prices in Illinois, 1970 to 1989.**

Land price change, percent

Real interest rate, percent

Changes in real land prices ..... Real interest rates

Source: Scott, 1989.

in the Corn Belt between 1972 and 1979 was 12 percent. The relationship between real interest rates and real land prices is illustrated in Figure 6.2.

Under these conditions, farm expansion made sense during the 1970s. The trend line was set. The incentives were too great. But keep in mind that most, if not all, of the trend line was based on the belief that in the 1980s monetary and tax policy would be the same as in the 1970s. At the time, few in the agricultural sector were concerned that policy would change. Bankers continued to lend money on the basis of their expectations that land prices would remain a good hedge against inflation.

In the fall of 1979, however, monetary policy did change. The Federal Reserve, in isolation, decided that inflation was public enemy number one, so the money supply was restricted. *No other single policy decision has caused more problems for the U.S. farming sector.* The sector was more indebted than at any other time in U.S. history and, unavoidably, it took an enormous economic hit.

Restricting the money supply resulted in a higher cost of money. Interest rates increased and inflation was reduced. Thus the real rate of interest was hit from both sides. The Federal Land Bank interest rate averaged around 12 percent from 1981 to 1986, while inflation averaged 4.7 percent. The real rate of interest was 7.3 percent versus the 1.3 percent for 1972 to 1980. Furthermore, as the dollar rose relative to other currencies, agricultural prices fell (see Chapter 7). On the heels of the increase in real and nominal interest rates and the decline in output prices, land prices declined in both nominal and real terms (Figure 6.2).

In short, changes in monetary policy during the 1980s brought about the massive devaluation of land, the major reason for the financial stress of that period. Farmers who had debt loads tied to higher land values suddenly found that their debts exceeded the value of their assets. Nervous creditors began foreclosing on loans. Particularly hard hit were farmers who had used the strategy of servicing debt by refinancing land loans during the 1970s. These farmers and their bankers had been confident that inflation would continue and land prices would not decline. They were wrong.

### The Cash-Flow Problem

Inflation is among the most serious sources of farm problems, and U.S. farmers should applaud the attempt to control it. Inflation and inflation expectations create a "cash-flow problem." The essence of the problem is captured by the old saying, "Farming is the only occupation where you can lose money every year and retire a millionaire." If a farmer can sell the farm, of course.

The cash-flow problem arises because of two components to earnings from ownership of land. The earnings that come from farming generate income that flows to the farmer. But people also expect land prices to increase at levels equal to or greater than inflation because land, like gold, is an asset in fixed supply. However, these returns as cash flow are not realized until the land is sold. Indeed, during hard times, they may never be forthcoming. When farmers borrow to buy land, the cash-flow problem rears its ugly head unless earnings are high (Box 6.3). Although earnings in any particular year may not be enough to service the debt, the value of land can cover the loan. It does so at a high price, however. In effect, the cash needed to service the debt increases the cost of acquiring land far beyond initial purchase price.

### Price and Output Risk

Both the cash-flow and the asset-value-risk problems are long-run problems for farmers. Much of U.S. production is now structured

Box 6.3.

# The Cash-Flow Problem

During times of rising prices, investors in land assume that land prices will increase at a rate at least equal to inflation. But interest rates also reflect the rate of inflation. This creates the cash-flow problem. To illustrate the problem, let's assume that the inflation-free rate of interest (or the real rate of interest) is 3 percent and that is also the rate of return from farming.

A farmer having $100,000 in assets who wants to buy 200 acres of land at $1,500 per acre would have to borrow $300,000. The resulting $400,000 in assets generates a 3 percent rate of return each year, or $12,000. If there is no inflation, the interest rate will be close to 3 percent. In this situation, the debt service on the $300,000 would be $9,000. Thus, without inflation, the farmer has $3,000 as a rate of return on his $100,000 of equity.

With inflation of 6 percent, the interest rate is 9 percent (3 + 6). The rate of return from farming can still be 3 percent, yielding the same $12,000 per year. But the debt service is now $27,000, so there is a shortage of cash equal to $27,000 minus $12,000, or $15,000. In other words, the farmer is short $15,000 when it comes time to pay on the loan.

Still, the farmer may not have made a bad decision. The $400,000 of assets will inflate at 6 percent each year ($24,000). The $24,000 in capital gains offsets the minus $15,000 in cash flow by $9,000, or 9 percent of the original $100,000 in equity. Markets are working, but it is clear that only farmers who have other income to offset these negative cash flows can expand under these conditions. Their wealth position is increasing, but the only way to take advantage of that wealth is to sell assets or to borrow against them.

Adapted from Bruce Gardner, *The Governing of Agriculture*, 1981.

around them. Price and output risk can exacerbate both of these problems. Farmers who use credit are exposed to more risk from the cash-flow and asset-value-risk environment. For these farmers, the price and output risk of every growing season are important. Farming is a risky business. Prices can change dramatically within the growing season, and production is always vulnerable to climate and weather changes.

Dramatic crop losses and price declines create serious cash-flow problems, even though a farmer may have planned well. In fact, farmers generally do plan very well for risk and uncertainty. Some, however, may need help to do so more effectively. Planning becomes critical when the asset-risk and cash-flow problems created by inflation and changes in inflation expectations are at their worst. Such conditions also intensify the need for ways that farmers can manage their price and output risk within the growing season.

## Rethinking the Farm Problem and Farm Policy

*Stability of income and asset values, then, is the real farm problem, at least from a production standpoint.*

Stability of income and asset values, then, is the real farm problem, at least from a production standpoint. Recognizing this structural reality of U.S. agriculture is absolutely necessary to rethinking policy. The further polarization of farms into small and large types, as we discussed in Chapters 3 and 4, points to two different categories of farms that face different problems and opportunities. Thus policy should be sensitive to their different needs. On the one hand, most small farms have off-farm income to help them stabilize family income. To help these farmers, public policy needs to be concerned with off-farm job opportunities. On the other hand, large farms need a stable macroeconomic environment and risk management opportunities that will not be bid into asset values.

### Small Farms Need Rural Development

Recent trends highlight the importance of off-farm work for small and relatively stable farms. Operators of mid-sized farms appear to have two choices: to increase their farm size and become more commercial, or to find off-farm work and downsize their operations.

Those who identify the farm problem as an income problem for the small farms should examine the numbers. In 1988, farms with less than $40,000 in gross farm sales had farm incomes that averaged only $2,000 per farm. However, their off-farm income averaged over $26,000. These farms account for more than 70 percent of the 2.2 million farms in the United States.

Quite obviously, policymakers, following the logic we outlined earlier in the book, must rethink the interdependence of farm and rural communities. As explained in Chapter 3, rural communities depend less and less on farms. Given social changes, farms depend more on rural communities than rural communities depend on farms. These new linkages between farms and rural communities intensify the need to diversify job opportunities and rural development. Traditional farm programs simply do not contribute meaningfully to rural well-being for most residents.

### Commercial Farms Need a Stable Macroeconomy and Risk-Management Tools

If the farm problem for large commercial farms is not the level of income, but rather the variability of income and a cash-flow problem, then current farm policy is misdirected. Any policy that distributes so much of its outlay to but a few producers should be reexamined. Let us reinforce an earlier observation: 15 percent of the farmers receive 73 percent of all benefits, and landowners benefit more than operators of farms. This inequity is difficult to rationalize.

*Current government policies also introduce risk and uncertainty for U.S. farmers.*

And other policy problems exist. Because government actions designed to support farm incomes generally are bid into land prices, changes in government actions can adversely affect land prices. To the extent that a change in any of several government farm and financial policies directly affects a farmer's balance sheet, this risk threatens farm viability. In short, current government policies also introduce risk and uncertainty for U.S. farmers.

Despite subsidies, large commercial farmers are not getting what they need from agricultural policy. They need a stable macroeconomic environment, including low and stable inflation and a competitive dollar exchange rate. But it has become difficult for the U.S. government to achieve stability in isolation. The large, unrestricted flows of capital among countries make macroeconomic management a matter for international coordination. For their own good, farmers and other internationally directed businesses should promote efforts for international macroeconomic management. Unfortunately, doing so is not on the immediate horizon, because political support in the United States and abroad is limited.

As agricultural markets have become international, risk and uncertainty in farming will continue. New forms of assistance are needed to help farmers cope with risk, forms that will not lead to higher land prices or rely heavily on taxpayer dollars. Mechanisms where farmers share the cost of managing risk are one alternative.

During the 1980s, Congress took two steps towards these policy goals. Legislators significantly changed federal multiple-peril crop insurance and reopened options trading in the futures market. Both of these steps require farmers to pay for risk protection. In principle, crop insurance and options markets work in the same way. In both, farmers must pay to protect themselves against low yields and prices. Because farmers pay for this risk protection, land prices are affected less than they are by direct income transfer programs.

Other policy options exist, though. Another possible risk-management tool is area revenue insurance. Government policymakers could design a program to provide insurance to all farmers at the same payment rate in a given locale. Payments would be based on the degree to which the area revenue for a specific crop is below average. For example, farmers would select the level of protection desired, such as dollars per acre. If revenues for an area, such as a county, drop below a specified level, the percentage of the drop would be used with the dollar coverage purchased to calculate an indemnity payment for each local producer. Because farmers would pay a premium equal to the risk protection offered, asset values would not be influenced by such a plan. The plan would also provide more stability for all of U.S. agriculture, and it would be cheaper to administer than individualized insurance programs.

Several market mechanisms could help farmers cope with risk and uncertainty in a way that is not self-defeating. That is, they would not bid benefits into land prices. For example, can farmers begin to use longer run futures market strategies to hedge against adverse movements in interest rates? Successful farmers of the future will probably use mechanisms like this to protect themselves against risk. But to use such mechanisms, farmers must be innovative and educated. Land-grant universities and extension services must foster innovation by training farmers to manage their risks. This is an important service to an old constituency. These educational institutions need to be innovative, as we have emphasized in Chapters 3, 8, and 9.

## A Final Caveat

Socialist economies have recently rediscovered the role that prices play in communicating important signals to producers. There is no such thing as a "fair" price in a market economy. Yet, ironically, *notions of fairness persist* in setting farm price and income supports in the United States. These government payments to farmers have been self-defeating because they raise costs of production and shut agriculture out of international markets.

Farmers face significantly more risk than other business persons do, and not just because of variations in yield. Changes in the macroeconomic environment can drastically alter land values, with disastrous consequences for the profitability of farming. Commercial farmers, like other internationally oriented producers, need a stable macroeconomy. In addition, farmers need risk management tools to deal with the unique uncertainties of farming. Most of these tools can be provided through market mechanisms. If we go on using farm price and income supports to solve problems created by changes in the macroeconomy and international markets, we can only recreate old wrongs.

# 7 Never Forget That U.S. Agriculture Depends on the World Economy

In his essay "Self-Reliance," Ralph Waldo Emerson used examples from farming and pioneering to show that people must accept full responsibility for their lives if they are to realize their potential. By accepting responsibility for failures, even those caused by acts of God, farmers internalize an ethic of creative industry, of seeking new ways for doing things when their plans are unsuccessful. In Emerson's view, self-reliant people see themselves as fully accountable for their future, without regard to forces beyond their control. As in many of Emerson's essays, the trait of self-reliance comes easier for yeoman farmers than for others, and their example must serve as a lesson for us all.

Emerson discusses how farmers achieve self-reliance through producing a wide variety of foods and other products they need for their livelihood. Farmers living on a frontier had to produce these items to survive; failure to do so was one's own fault. Many readers of "Self-Reliance" may have regarded farmers' ability to produce all that they needed (and nothing more) as the central message of the essay, rather than as an illustration of a larger point. A diluted notion of self-reliance would simply mean that the farm or the farm community is a closed system, a self-sufficient group of people with no dependence on the larger world.

Equating self-reliance with self-sufficiency was, no doubt, what George Washington was getting at when he warned the American nation to avoid foreign entanglements. The result was a world view that, even into the 20th century, promoted insularity and protectionism. Isolationism, or closing the nation's gates, was a main theme of political debates for at least two thirds of the 19th century. Ironically, in seeing foreigners as external threats to American order, this world view is almost the opposite of the open-ended pursuit of autonomy that was the true message of "Self-Reliance."

The value placed on economic self-sufficiency surfaces in farm interests as a mistrust of international markets. Since the Great Depression, agricultural populists have tended to be isolationists. In their view, relying on foreigners for income is foolish, and tailoring production or policy to promote trade is not in the national interest. Some advocates of sustainable agriculture express similar views, because they see international trade as interfering with farm income support and environmental goals.

At the other end of the political spectrum, commodity groups have taken an aggressive, mercantilist approach to exports. Free trade is okay as long as it is practiced by other nations, particularly those that buy our products. Direct export subsidies and subsidized marketing campaigns for agricultural products are part of agriculture's special entitlement to public resources. Promoting agricultural exports is added to the long list of policy objectives for domestic farm programs.

Although commercial interests tend to dominate the agricultural policy debates, the populist antitrade sentiment is still used to justify the basic mechanisms of farm policy. That irony shows how the many and quite divergent interests structure bits of agricultural policy, dragging into the process inconsistencies in direction. In this chapter we explore the inherent contradictions between the economic realities of world markets and the assumptions behind farm programs.

## Why Is Agricultural Trade Important?

The reality is that the self-sufficient, mixed farm of yesteryear no longer characterizes agricultural production in the United States. Production has become highly specialized, with most farms producing only a few commodities. As transportation has become less costly and as international demand has grown, U.S. agriculture has changed in response to new market opportunities. Yet many farm interest groups still demand that the United States ignore one or another aspect of world markets when making domestic farm policy. These groups have not recognized four economic facts of life:

- Exports are the primary growth point for agricultural product demand.
- Export demand is the major determinant of farm incomes and farm program expenditures.
- Government policies other than the farm bill are very important in determining agriculture's ability to export.
- Supply-control and price-support programs can damage U.S. farmers' ability to export.

## Exports: The Primary Growth Point
## for Agricultural Product Demand

As consumer incomes grow, demand for food grows, but more slowly. Per capita demand for most foods eventually reaches a saturation point. Because the United States is a wealthy country, domestic demand for food grows only about as fast as the population. Income growth brings demand for higher quality and more processed food, rather than greater quantity. Demand for individual products may grow faster than the population, but usually these products are simply replacing other agricultural goods in the diet. New sources of demand are rare.

The primary growth point for agricultural demand during the last twenty years has been export demand. For example, our domestic demand for wheat, rice, and coarse grains grew at 1.4 percent annually from 1960 to 1989, but exports of these crops grew much faster at 4.3 percent. Exports have taken an increasing portion of all agricultural sales since 1970 (Figure 7.1). An exception to the upward trend was the export slump of the early 1980s. Exports recovered in the late 1980s, and by 1990, trade accounted for about one quarter of agricultural output and one third of acres harvested.

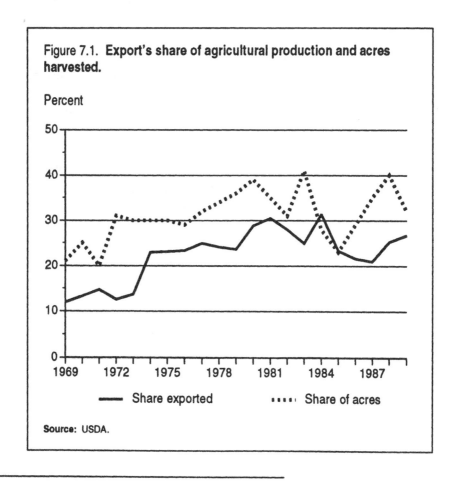

Figure 7.1. **Export's share of agricultural production and acres harvested.**

Percent

— Share exported    ••••• Share of acres

Source: USDA.

Our dependence on exports is even more pronounced for feed and food grains, which make up the bulk of government program payments. During the late 1980s, these commodities together accounted for 70 percent of program payments. Food grain producers sold two thirds of their production overseas; feed grain producers relied on the international market to take more than one fourth of their production.

## Export Demand: A Major Determinant of Farm Incomes and Program Expenditures

As our economy grows, the demand for food declines relative to the demand for other goods. Resources thus shift out of agricultural production into manufacturing and services. This process of shifting resources out of agriculture has been slowed because we can export to other countries where demand for food is still growing. Without exports, the United States would produce less food with fewer resources. Export sales have slowed the inevitable decline of agriculture's importance in the economy. When export demand has been strong, it has reduced the taxpayers' cost of supporting farm incomes.

> When export demand has been strong, it has reduced the taxpayers' cost of supporting farm incomes.

Because exports are the primary growth point in demand, they are a major determinant of farm prices and incomes. In economic jargon, exports provide the marginal change in demand that determines price. As Willard Cochrane has pointed out, farm income followed trends in agricultural exports during this century. But this was not true in the 1930s and again in the 1980s, because government support raised farm incomes in spite of declining exports. During the last twenty years, government payments have been inversely related to the level of export sales (Figure 7.2). The export boom of the 1970s boosted farm income, and government program costs remained low, at around $2 billion to $3 billion per year. Agricultural exports then plunged from a peak of $44 billion in 1980 to $26 billion in 1986, while government program expenditures soared from $4 billion to over $20 billion. Since 1986, farm programs have cost about $15 billion per year on average. These greater government outlays preserved gross farm income at around $165 billion per year.

## Government Nonfarm Policies: An Important Determinant of Agriculture's Ability to Export

Emerson could not have foreseen that the following three factors would determine U.S. farmers' ability to export:

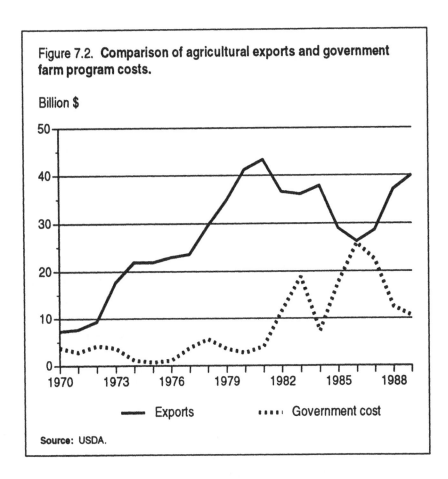

Figure 7.2. **Comparison of agricultural exports and government farm program costs.**

Billion $

1970   1973   1976   1979   1982   1985   1988

—— Exports    ••••• Government cost

Source: USDA.

1. *The U.S. dollar exchange rate:* The value of U.S. currency determines how much our customers have to pay for our products. As the number of Japanese yen needed to buy one U.S. dollar increases, so does the price of U.S. corn for a Japanese buyer.
2. *Economic growth in the third world:* Consumers in less developed countries have pent-up demand for more and better food, held back by low incomes. As their incomes grow, their demand for food, and hence food and feed imports, increases.
3. *Protection of agriculture in other industrial countries:* Other high income countries have protected farm incomes by restricting imports of agricultural products and by subsidizing exports.

If Emerson had foreseen these factors, and if the difficulties of influencing them had been understood, he might well have been an isolationist rather than just a romantic. The surprising importance of these factors has been demonstrated only too well by the events of the last twenty years. The boom and bust cycle of agricultural exports between 1972 and 1985 shows how closely agriculture's fortunes are tied to international developments and U.S. macroeconomic policies.

During the 1970s, the value of the U.S. dollar gradually fell. Exports were then cheaper to foreign buyers. At the same time, income growth overseas and changes in Soviet policy led to greater meat consumption abroad. Many countries imported U.S. feed grains for their expanding livestock production. Income growth was most rapid in less developed economies, and these countries took an increasing share of U.S. feedstuffs exports. Agricultural exports boomed, from less than $10 billion in 1972 to $44 billion in 1980 (Figure 7.3). The economic environment changed drastically in the 1980s, leading to a sharp decline in the volume and value of U.S. exports. Boom turned to bust.

What happened? First, the exchange rate raised the price of U.S. farm commodities to overseas buyers. For example, the price of soybeans in German marks increased 24 percent between 1980 and 1984. As the foreign currency prices of U.S. agricultural exports climbed, demand for U.S. farm products fell.

Second, the global recession of the early 1980s slowed the growth in demand for agricultural exports (Table 7.1). The debt crisis in developing countries also slowed their growth and ability to import.

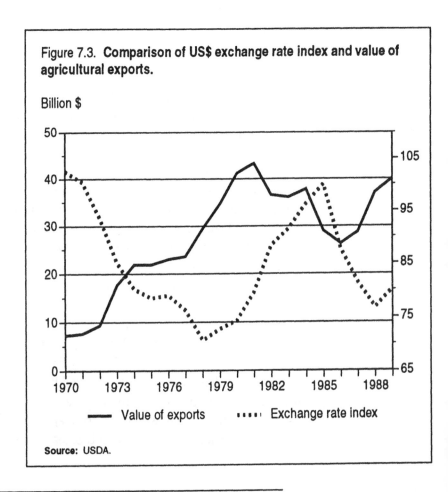

Figure 7.3. **Comparison of US$ exchange rate index and value of agricultural exports.**

Billion $

—— Value of exports    ⋯⋯ Exchange rate index

**Source:** USDA.

Table 7.1.
**Agricultural Trade as Related to Economic Growth (Average Annual Growth Rate in Percent)**

| | | GNP per Capita Growth in: | |
| --- | --- | --- | --- |
| Period | World Wheat and Coarse Grain Trade | Low and Middle Income Countries | All Market Economies |
| 1973-80 | 5.5 | 2.6 | 1.5 |
| 1980-86 | –1.6 | 1.5 | 0.9 |
| 1987-89 | 3.2 | 2.4 | 2.1 |

**Source:** USDA; World Bank.

Third, foreign protectionism also reduced markets for U.S. products. A prime example was loss of the European market for grain exports due to the European Community's Common Agricultural Policy (CAP). The CAP provided European farmers with high support prices insulated from world markets. European grain producers responded with increased production. Eventually, Europe no longer needed to import grains. In fact, the EC began to subsidize the export of surplus production onto world markets. The EC shift from grain importer to exporter reduced demand and increased competition for U.S. agricultural products.

U.S. agricultural exports fell sharply in the early 1980s (Figure 7.3), but the bust was not permanent. Exports have recovered gradually since 1986, as the value of the U.S. dollar fell and the 1985 farm bill lowered support prices. Recovery of the world economy also boosted demand for agricultural exports.

Because events throughout the world influence export demand, they often seem beyond the reach of U.S. influence. But that is only partly true. U.S. government policies play a role in shaping the economic environment for exports. Our fiscal and monetary policies, of course, influence the exchange rate. To encourage demand, U.S. trade policies can foster economic growth in less developed countries by allowing their exports to enter U.S. markets. The United States can also take the lead, as it has since 1986, in pushing for liberalization of agricultural trade. The latter two policy goals depend on luck and skill, in that both require cooperation from other nations.

## Supply-Control and Price-Support Programs: Damaging to Farmers' Ability to Export

Since 1933, U.S. agricultural policy has been founded on the assumption that supply control and market intervention will raise farm prices and incomes (see Chapter 6). But that assumption is logically flawed. Policies attempt to support prices either by limiting domestic production or by government purchases at a floor price. If the United States neither exported nor imported agricultural products, then farm programs *could* determine domestic supply and prices. But with international trade, supply control is ineffective in setting domestic prices, and market intervention is very costly.

When a commodity is imported, trade must be limited to make domestic supply control work its desired end. To support prices of dairy, sugar, tobacco, and peanuts, the United States imposes import quotas for these products (see Box 7.1). It is then feasible to raise domestic prices either by controlling production, as is done for tobacco and peanuts, or by government purchases to defend the support price, as is done for dairy.

Much of our agricultural production could be competitive in export markets. But for exported commodities, supply control or price supports cannot maintain higher incomes. For example, restricting supply initially raises prices, which reduces market demand. Our government must then purchase and dispose of the surplus production. Furthermore, higher U.S. prices make U.S. exports less competitive. Over time, our share of the world market becomes progressively smaller as other countries increase their production. As export demand shrinks, then domestic production must be restricted even further. Either that, or the government must pay to dispose of the larger surplus that cannot be sold in the market.

*Farm programs have reduced U.S. agriculture's ability to compete in world markets more than once since 1933. A recent example of highly counterproductive market intervention is found in the 1981 farm bill.*

Farm programs have reduced U.S. agriculture's ability to compete in world markets more than once since 1933. A recent example of highly counterproductive market intervention is found in the 1981 farm bill. Market support prices, or loan rates, for commodities were raised and locked into place because continued high inflation and world demand were expected. When inflation slowed and export demand declined, enormous downward pressure was put on commodity prices. But policy had locked into place high support prices, and our competitors in world markets were able to sell well under U.S. prices. As a result, the U.S. market share of world trade declined, the volume of U.S. exports dropped, and surplus grain entered government stocks. That was how the 1981 farm

Box 7.1.

# The GATT and Agriculture

The General Agreement on Tariffs and Trade (GATT) was established in 1947. Its primary purpose is to promote free trade and to reduce trade barriers. Countries signing the agreement recognized that protectionist policies during the worldwide depression of the 1930s had been disastrous for economic growth. GATT was to lay the foundation for a new era. Since 1947, average tariffs in industrial countries have fallen from 40 percent to 5 percent, and trade in manufactured goods has grown 20 times.

GATT rules have three principal elements:

- Each country will treat all other members in the same fashion. (This is the nondiscrimination principle.)
- Each country makes a commitment to observe negotiated tariffs.
- Quotas on imports or exports are prohibited because they limit competition more severely than tariffs do.

In the original GATT articles, agriculture was recognized as having "special" status, and possible exemption from the above principles. The third rule would have altered the operation of some U.S. agricultural price-support programs, such as those for dairy and sugar. In 1955, the United States received a waiver of GATT rules in order to continue import quotas on agricultural goods. In later years, the Europeans became increasingly interested in excluding agriculture from GATT rules because many features of their Common Agricultural Policy would probably not be permitted.

The Uruguay Round of trade negotiations under GATT started in 1986. It was the eighth round of talks seeking reductions in trading barriers. For the first time, agricultural trade was a major area for negotiation. Trade in services and intellectual property rights also were included for the first time. The United States and the Cairns Group, a coalition of fourteen agricultural exporting nations, pushed for substantial reductions in agricultural trade barriers. The talks were to have concluded in December 1990, but collapsed when the EC disagreed with agricultural exporters, including the United States, about agriculture. Efforts to revive the Uruguay Round have depended on reaching an agreement about how to reduce agricultural trade barriers.

bill contributed to the poor export performance of U.S. agriculture in the early 1980s.

The 1985 farm bill was an attempted correction. Policymakers reduced market support prices or loan rates in a bid to make our agricultural exports more competitive. To maintain farm income, direct payments to farmers were kept at relatively high levels. The explicit goal of the 1985 bill was a fire sale of exports to regain U.S. world market dominance. However, the bill still contained major elements of supply control. Set-asides of farmland were very high as percentages of total crop area. And the new Conservation Reserve Program took land out of production, achieving a small amount of conservation and a large amount of supply control. According to USDA (August 1990), this foregone production was matched by increased exports from competitors in the world market. Idling land thus helped competitors gain market share.

Although much lip service has been paid to keeping agricultural exports competitive, arguments for supply control and price supports are hard to eliminate. Both were major elements in 1990 farm bill debates. Despite the lessons of the 1980s, raising market support prices was still a dominant goal of commodity organizations.

Rather than promote more competitive production, agricultural interests have called for targeted programs to boost exports. A recent example is the Export Enhancement Program (EEP) of the 1985 and 1990 farm bills. This program uses surplus stocks, mainly of wheat, to provide in-kind subsidies for exports. The rationale is that it counters competition from the EC's subsidized wheat exports. Robert Paarlberg has called EEP "mysteriously popular." As he noted though,

> EEP hasn't added much to foreign sales because 9 out of 10 EEP bonus bushels simply displace sales that would have been made anyway. It transfers more benefit to importing countries than to U.S. farmers. Instead of working against the EC, EEP is mostly working to antagonize Australia and Canada—our natural allies against EC in the GATT. (p. 14)

The problem with targeted programs like EEP is that they try to counter the overall effects of illogical U.S. policy. Such programs cannot address the root causes of export slumps.

An unwillingness to rely on international markets or to formulate a broad-based trade strategy has been costly. Taxpayers have lost as they have footed the bill for the old ideas of farm income and price support *and* for the new ideas of export subsidies. Consumers of dairy and sugar have lost by paying higher prices due to import quotas. Agriculture has gained the short-term benefits of price and income support at the expense of lost world market opportunities.

# What Kind of Policies Make Sense in an Interdependent World?

In some significant ways, a mistrust of foreign markets is not misplaced. International markets do change rapidly, and other countries' trade barriers are a real problem for U.S. agriculture. The boom and bust cycle of exports have wreaked havoc with U.S. farm incomes and asset values during the last two decades. Rapid increases in domestic prices caused by export growth in the 1970s also brought consumer protests. The causes of price and trade fluctuations often seem beyond the influence of domestic farm interests. So why not avoid the world market altogether?

U.S. agriculture cannot turn its back on international markets. To do so would mean dismantling a large part of the agricultural sector. Therefore, to keep U.S. agriculture going, policymakers must abandon the traditional assumptions behind current farm programs. They need to think more broadly about policies to make sure that agriculture has the most favorable prospects in export markets. The economic facts of life have several implications for policy.

> *U.S. agriculture cannot turn its back on international markets. To do so would mean dismantling a large part of the agricultural sector.*

## Farm Programs Should Not Work Against Competitiveness

Farm interests have been shooting themselves in the foot for many years by trying to manipulate prices. Raising prices, either through supply control or government purchases, will only make the United States less competitive in world markets. Farm income support must be administered so that farmers can respond to world market signals in their production decisions.

In this regard, the last two pieces of farm legislation set important precedents. Program yields were frozen by the 1985 Food Security Act at the levels of the mid-1980s. In the 1990 Act, the number of program acres eligible for government program payments was reduced. This means that, for the last bushel or two produced, farm income support has been severed from production. Any increases in production must now be profitable at world prices. Thus farmers will increase or lower production of program crops, or shift those acres ineligible for program benefits to other crops, in response to changes in world prices.

## Export Subsidies and Promotions Are Costly in Relation to Results

When macroeconomic conditions and general farm policy interfere with agriculture's ability to export, it is unwise to counteract with direct subsidies. The costs are too high. For example, the average subsidy for wheat exports under the Export Enhancement Program from

1985 to 1989 was 26 percent of sales value. In other words, for every dollar of wheat exported, U.S. taxpayers paid 26 cents and foreign buyers paid 74 cents. Moreover, trying to boost demand by throwing tax dollars at foreign buyers does little to foster long-run growth in export markets. They, like U.S. automobile consumers, learn to wait for the next rebate or buy from foreign competitors.

By driving down world prices, export subsidies and promotions can also harm efforts by the less developed countries (LDCs) to nurture their own agricultural production. In this regard, the United States and the European Community use their considerable economic power to hurt the "little guy." Besides being unfair, this tactic can also work against U.S. self-interest in the long run by constraining growth in LDCs. This leads to our next point.

---

**Box 7.2.**

## Environment and Food Safety Concerns: New Challenges for Resolving International Trade Disputes

The growing public demand for a better environment and a safer food supply surfaced in the 1991 debate over renewing negotiating authority for the U.S. trade representative. Consumer groups were concerned that international trade rules would take precedence over domestic regulations regarding food safety. Environmentalists also opposed trade negotiations in the interest of maintaining domestic autonomy for environmental policy. They were afraid that trade rules could limit domestic environmental regulation of agriculture.

These new public concerns pose two questions for trade negotiators who want to reduce trade barriers. First, when does an environmental or food safety regulation reflect a legitimate concern, and when is it a cleverly disguised barrier to trade? Second, who will be responsible for making this distinction? The answer to the second question will prove to be difficult, because few countries are willing to submit to international arbitration for these issues.

A few guiding principles seem clear. First, different countries will undoubtedly have different costs and benefits from imposing environmental regulations. Thus it is unreasonable to impose one set of standards on all countries. Second, when countries do impose a food safety standard or an environmental regulation, it should be applied in the same way to both imports and domestic production. Third, it may be easier to arbitrate food safety disputes than it is to arbitrate disputes over environmental regulations. Food imports can be tested at the border for harmful substances. But the intent of environmental regulations, as well as their impact on trade, will be more difficult to assess.

---

### U.S. Trade and Debt Policies Must Encourage
### Third World Economic Growth

Demand for exports to LDCs has the best potential for U.S. trade growth. These nations want, but cannot yet pay for, more and better food. Many of them are still struggling to recover from the 1982 recession and the burden of foreign debt. As long as economic growth in LDCs is slow, agricultural import demand will also be slow. Some farm groups have recognized this fact by supporting open markets for third world imports and liberal terms for debt relief.

### It's Important to Negotiate Away Trade Barriers,
### Even Though It's Frustrating and Unglamorous

The future of agricultural exports depends on the reduction of trade barriers. In the industrial nations, agricultural trade is limited by the high level of government intervention in agriculture. New concerns about food safety and the environment may raise still more barriers to agricultural trade (see Box 7.2). In the newly industrializing countries, where demand is growing, trade barriers can easily be put in place. One example is South Korea. That nation has emerged as a major buyer of U.S. agricultural exports during the last fifteen years, but it has gradually put policies in place to restrict growth in imports.

Negotiations under the GATT to lower agricultural trade barriers dragged on with little result (see Box 7.1). Still, a GATT agreement on agriculture is worthwhile. Bringing agriculture under international trade rules would set the stage for future negotiations. GATT at least provides a framework for negotiating future reductions in trade barriers and prevents new importers from raising barriers. Much of U.S. agriculture would benefit from more open world trade.

### Recognize That an Uncertain Macroeconomy Is Here to Stay

Agriculture is particularly vulnerable to a high value of the dollar because the sector relies so heavily on export demand. High interest rates also hurt agriculture because they raise production costs (see Box 7.3). Changes in the interest rate, the inflation rate, and the exchange rate, as noted in Chapter 6, have caused major adjustments with heavy costs to agriculture. But the United States will have less control over macroeconomic policy in the future than we had in the past. Not much can be done about this, except to plan for it. Floating exchange rates, a large open international capital market, and an intractable federal budget deficit all limit the ability of the U.S. government to manage our economy. These changes impose increasing uncertainties on agriculture and make farming an even riskier business than it has been.

---

Box 7.3.

## Why Macroeconomic Policies Are Crucial to Agriculture

U.S. monetary and fiscal policies affect both costs and returns in agriculture through the level of interest and exchange rates. Interest rates determine the cost of borrowing, and interest charges accounted for 16 percent of cash costs in agricultural production in 1987. Exchange rates determine the cost of U.S. commodities to foreign buyers. Because exports take a large share of U.S. agricultural sales, exchange rates affect the prices of commodities.

The real interest rate is the actual interest rate minus the rate of inflation. It represents the real cost of borrowing money. During the 1970s, real interest rates were low or negative, but they increased sharply during the early 1980s. These high interest rates were in part the result of increased government borrowing to finance the mounting federal budget deficit, which increased from under $50 billion in the 1970s to over $200 billion in 1985.

High real interest rates in the early 1980s attracted foreign capital investment into the United States. As foreign investors bought dollars, they bid up the exchange rate for U.S. dollars. Between 1979 and 1984, the real interest rate increased by 4 percentage points, and the value of the dollar increased by more than 40 percent.

The high real interest rate and the high dollar exchange rate squeezed farm income in the early 1980s. Farmers' interest expenses rose from 13 percent of cash costs in 1979 to 19 percent in 1984. As the prices of U.S. agricultural exports in foreign currencies climbed with the value of the dollar, demand for our exports fell. The value of U.S. agricultural exports plummeted from $44 billion in 1980 to $26 billion in 1986. As a result of rising costs and falling earnings, net farm income dropped sharply in the early 1980s.

Although U.S. farm income recovered in the late 1980s with a falling dollar exchange rate and large farm program payments, agriculture remains vulnerable to swings in the larger economy. Changes in the macroeconomy will surely occur and are likely to become more unpredictable. The interest rate, the inflation rate, and the exchange rate cannot all be controlled by U.S. fiscal and monetary policy, as they were in the past. The huge federal budget deficit will continue to put pressure on real interest rates. The large, open, international capital market will force the exchange rate to change with differences in real interest rates among countries. Thus the macroeconomic policies of other countries place a limit on how effective U.S. monetary policy can be. Changes in the macroeconomy will bring continued risks and uncertainties for agriculture.

## What Can We Conclude?

U.S. agricultural policy was conceived under the assumption that domestic markets are isolated from world influences. Policy still rides on that assumption. Its illogic first became apparent, however, when the United States was forced to seek GATT waivers for agricultural programs in the 1950s. During the last two decades, the influence of international markets on farm prices and incomes has become even more apparent. Exports are the most important determinant of farm income and, indirectly, of government payments to farmers. Yet policymaking has not caught up with the realities and new uncertainties of the world economy.

*Ignoring world markets could make us self-sufficient, but doing so will not make us self-reliant.*

In this unpredictable world, it is useful to remember Emerson's ethic of self-reliance. Ignoring world markets could make us self-sufficient, but doing so will not make us self-reliant. Just as acts of God required creative responses from yeoman farmers, so the challenges of international markets will require creative responses from modern agriculture. The alternative to self-reliance is a slow and painful decline.

# 8 Never Equate Good Farming with a Healthy Environment

Among the most tattered of agrarian myths is the belief that farmers place great value on stewardship and follow sound environmental practices accordingly. Stewardship values are firmly grounded in the triangle of farming, citizenship, and moral character discussed in Chapter 2, but these values are far from being well served. Nonfarmers have become increasingly aware that agricultural production can degrade the natural environment through soil erosion, pollution of surface water and groundwater, and destruction of wildlife habitat (see Box 8.1). Environmental advocates have raised serious questions about the strength of farmers' commitment to stewarding natural resources. We must therefore rethink agrarian beliefs about stewardship as we examine the relationship between agricultural policy and environmental quality.

The key idea in 18th and 19th century agrarian philosophy is that farming unifies self-interest and the public good. Jefferson emphasized that, if farmers have fixed assets in land, they probably will not encourage policies that could erode the stability of democratic government. Emerson stressed that farm work brings human needs into harmony with nature's cycles, making it natural for farmers to achieve self-realization through their life's work. Nonfarmers might achieve the virtues of citizenship and harmony with nature, but farm life is a moral ideal because these virtues are thought to be wedded to farmers' self-interests. Farmers do not need to resist the temptation of self-interest, for their self-interests are fully consistent with what altruism requires.

This theme is a highlight in Wendell Berry's writings on the importance of the family farm. Berry writes of how farm families experience the unity of nature within the diversity of their lives. As noted in

Box 8.1.

## Agricultural Production Can Harm the Environment

Through a number of mechanisms, agricultural production can lead to environmental damage, much of it felt beyond the farm. Policymakers and the public have recently turned their attention to some of these mechanisms:

- Pollution of surface water by eroded sediment from cropland
- Pollution of groundwater by nitrogen fertilizers, residues of agricultural pesticides, or nitrates from livestock manure
- Loss of wildlife habitat when wetlands are converted to agricultural uses
- Loss of ecological diversity through overgrazing of rangelands

The extent and importance of environmental damage from agriculture is widely contested, and in some cases largely unmeasured. For example, according to USDA, damage from cropland sediment costs off-farm users of surface water an estimated $2 billion to $6 billion every year (Ribaudo 1986). On the other hand, as Stephen Crutchfield notes, we know that more than one third of the United States has the *potential* for groundwater contamination from agricultural chemicals, but we know very little about the *actual* extent of contamination.

Yet two points remain clear. First, environmental damage from agriculture has become noticeable enough to attract public concern. Second, as environmental damage from other activities is increasingly regulated, the remaining damage from agriculture becomes more noticeable. These two factors ensure that environmental issues will remain part of agricultural policy debates.

Chapter 2, members of the family perform roles that are specialized by age and sex and that also define their place in the family's social order. The diversity of tasks is also reflected in the change of seasons—plowing in spring, nurturing the crops through summer months, harvesting in autumn, and then repairing tools and buildings in winter. The farm family is at one with nature not in the sense of pastoral bliss, but by unifying diverse economic, cultural, and environmental forces under the goals of family survival.

Berry suggests that those who live in the industrial world have lost the means for appreciating unity within diversity. In his view, modern society subsists on the conflict that arises when specialists follow their own detached and narrow self-interest. "Checks and balances," he writes, "are all applied externally, by opposition, never by self-restraint. . . . The good of the whole of Creation, the world and all its creatures together, is never a consideration because it is never thought of; our culture now simply lacks the means for thinking of it" (p. 22). As human beings become less reliant upon their own individual abilities to make flexible and ingenious responses to natural adversity, they lose the farm family's capacity to appreciate the importance of environmental harmony. For Berry, the transparent connection between nature's needs and the family's needs equates good farming practices with ecologically sound use of the rural environment.

It is important to see why this belief is a half-truth not only for the modern commercial farms that Berry and many environmentalists distrust, but for the traditional small farms as well. Certain aspects of environmental quality were and are closely aligned with farmers' production incentives. Farmers still have an interest in preserving soil fertility, for example. But other aspects of environmental quality are (and always have been) unconnected with farmers' interests and are even deeply inimical to them. Farmers as a group have done no more than nonfarmers to promote the preservation of endangered species. In fact, they have actively tried to wipe out large predator species and other "pests"—wolves, cougars, groundhogs—throughout the past two centuries. There is also an ever-expanding middle ground of issues where the link between farm interests and environmental quality is unsettled and contentious. For example, farmers themselves disagree about chemical use and water quality.

> *Some aspects of environmental quality are unconnected with farmers' interests and are even deeply inimical to them.*

Agrarian beliefs about good farming are not the only myths that influence our perception of environmental policy. Throughout our history, Americans have tended to understand nature as detached from human interaction. City dwellers who wish to "get back to nature" have sought wilderness areas, conservation preserves, and parks for their vacations, rather than farming areas. Biologists and ecologists have built their scientific models of homeostatic natural systems upon environments where humans, including those who farm, do not play a part. The result is that pristine, uncultivated places provide the dominant norm for what "the environment" should be like in the ideal.

Taken literally, pristine wilderness is a norm that agriculture, by definition, cannot hope to fulfill. It is not clear that environmental groups have ever attempted to merge the agrarian belief that good farming is perfect stewardship with this norm. However, uncritical observers of agriculture often seem to base their expectations upon precisely such an impossible set of goals. In this chapter, we refer to the mixture of agrarian stewardship and pristine nature myths as *the pastoral ideal.* Farm activists and their representatives in policy circles have often responded to environmental issues as if this self-contradictory agenda were the actual political program of environmentalists. As a result, farm advocates tend to retrench behind the myth of stewardship and to promote the pastoral ideal itself, despite its implied tension with any form of agricultural practice. Promoting the pastoral ideal has kept us from examining how farm advocates themselves might benefit from a fresh look at environmental quality.

Farmers, government agencies, and environmental groups all strongly value conservation. But because they see different uses for soil, water, and wildlife habitat, they have conflicting views about what counts as environmental quality. The public, too, fosters competing expectations. By believing in the contradictory values of pristine, uncultivated nature and farmers as natural stewards, the public relies on farmers' voluntary compliance with the highest environmental standards as they go about the necessary task of bending nature to their will. The related belief that those standards can be met without fundamentally adjusting farm practices accentuates reliance on farmers' good intent as opposed to government's need to regulate.

As a result of misconceptions about farming and the environment, we end up with contradictory and self-defeating government policies. On the one hand, farm programs encourage more production, which increases the use of erodible land and chemicals. On the other hand, farmers are required to comply with soil erosion standards and to refrain from plowing up wetlands in order to qualify for program benefits. The end result has not been satisfying either to farmers or to environmentalists, or for that matter to the Soil Conservation Service. Rethinking policy requires us to recognize how modern agriculture relates to the environment.

Furthermore, the contradictory values of farm stewardship and pristine nature are often in conflict with the equally powerful cultural value of material progress. Nature and material production are both highly valued in our society, though unevenly. This produces problems in adjusting society to the conditions of modern agriculture. Both belief systems produce confusing public policy discussions: farmers try to play off images of agrarianism and then champion technology, while preservationists struggle to hang on to the doomed value of a simplified production system. Few, if any, realistically see the consequences of agricultural industrialization.

## The Pastoral Ideal and Agricultural Policy

Government policy has given agriculture special treatment among economic sectors. For example, the "polluter pays" principle has

---

never been applied to agriculture. Rather, environmental regulation of agriculture has relied on volunteerism. The taxpayer has shared the cost of conservation measures.

Why has agriculture been treated differently? Don Paarlberg (pp. 5–13) has shown how agriculture's structure was different and farmers were seen as unique. Many reasons explain why family farms endure when other family businesses fail, and why agriculture is, on the whole, an industry with a highly individual ownership structure. Fredrick Buttel and Louis Swanson, two rural sociologists, propose that these same reasons also say a lot about resource management behavior in this sector.

First, agriculture is different because land is a key factor in production. The supply of land is finite, so land cannot be manufactured. Second, agricultural commodities are necessities. Hence demand for them tends to grow slowly and to adjust slowly to price changes. The result is frequent episodes of market saturation, overproduction crises, and product price volatility. Third, these economic risks are exacerbated by the risks associated with climatic and other natural vagaries. And, fourth, the tie of agriculture to the land, and hence to vast reaches of geographical space, limits the economies of size found in horizontal and vertical integration of production activities in other industries.

Agricultural production clearly has become more concentrated in fewer but larger farms, but in the United States it is still within family farms. This socioeconomic basis for the structure of the family farm has two interesting implications for managing agricultural resources. First, the very characteristics of agriculture that lead to degradation of soil and water resources—product price instability, the cost-price squeeze, and the high-risk, short-term planning horizons—are the same characteristics that lead to family ownership of farming.

Second, prosperity does not breed stewardship for individual farmers. The *prosperous family farmer* constitutes much of the historic imagery of a long-term solution to soil and water degradation problems. The conservation behavior of U.S. farmers, so the argument goes, would improve significantly if we had many prosperous family farmers with enough capital to invest in conservation, longer term planning, and greater certainty of passing on the farm within the family. But the notion of a "prosperous" family farmer is to some degree a contradiction in terms, at least when we compare it with a prosperous family that operates a successful bank or an automobile dealership. The structural reasons for the continued existence of the family farm also create, almost by definition, the conditions leading to low and variable returns from agricultural investment.

Furthermore, farmers do not bear all of the costs of environmental degradation. Even prosperous family farmers who conserve their own productive resources will let others bear costs outside the farm. Society would probably be far less tolerant of destruction of wildlife habitat and chemical pollution of groundwater if farming were more like industry. The image of the struggling, financially strapped family farmer who should not be regulated into bankruptcy prevents mandatory regulation of agricultural practices under the "polluter pays" principle.

Yet despite beliefs about farm stewardship, resource degradation persists, as illustrated by changes made in policy assumptions about soil conservation. In the 1930s, three assumptions with obvious stewardship roots guided resource conservation. First, it was believed that soil erosion existed because of low farm prices caused by the Depression. Once the Depression was over and prices were higher, farmers would practice conservation. Second, it was supposed that farmers paid the primary cost of soil erosion as the productivity of the soil declined with use, so there are built-in incentives for farmers to accept the costs of conserving soil. And, third, it was determined that, given the first two assumptions, any particular program could be voluntary and still effective.

Each of these assumptions has been overturned. First, as Earl Heady and Carl Allen showed, farmers practiced limited conservation even when commodity prices were high. The second assumption was knocked down during the early 1980s, when USDA showed that the primary damage from soil erosion and other agricultural pollution occurred beyond the farm lot. The pollution of surface water and groundwater, the silting of ditches, the loss of wildlife habitat, and the draining of wetlands were primarily a cost to the nonfarm sector. The third assumption, that volunteerism was a viable policy option, was abandoned with the conservation compliance provisions of the 1985 Food Security Act. That legislation recognized that farmers need specific incentives to practice conservation, in this case, continued government entitlements.

## Where Pastoralism Fails: Why Belief and Behavior Diverge

The contradiction between widely held stewardship values and continuing environmental degradation from agriculture stems from the following three causes, none of which is unique to agriculture:

- Actions that seem insignificant to an individual can have a large environmental impact when added up.
- Social institutions and practices have not caught up with the

broadened agenda of environmental concern.
- Government policies sometimes encourage environmental degradation.

Each of these causes has affected the relationship between agriculture and the environment.

### Actions That Collectively Can Have Large Environmental Impacts

Farmers seldom realize when agriculture causes damage to the environment. An example is wildlife habitat destruction. One farmer drains a pothole to increase production and reduce the cost of plowing around it. But if all farmers in the northern prairies were to drain their potholes, they would eliminate breeding grounds for a great many of North America's waterfowl. Another example, noted by Pierre Crosson, is the high cost of off-farm damage from soil erosion, which USDA and the Conservation Foundation estimate to be three times greater than on-farm damage to productivity. The major off-farm costs, largely due to siltation in bodies of water, stem from reduced water recreation, increased flood damage, reduced water storage capacity, and more maintenance for waterways. None of these costs is directly evident to the farmers upstream.

Farmers are understandably reluctant to reduce environmental degradation if it means lower profits. Reducing off-farm damage usually costs farmers more without boosting farm profits. Still, these economic realities should not devalue stewardship, nor should acknowledging the potential conflict between farmers' individual property rights and the public good. Rather, these are necessary first steps for coming to grips with environmental issues.

### Social Institutions and Practices That Have Not Yet Caught Up with Some Environmental Concerns

Environmental policy has evolved over time to include goals that benefit farmers less directly. In the 1930s, soil erosion was the focus of agricultural conservation policy. Since then, agricultural practices have changed. Chemical use has become common: the amount of pesticides more than doubled between 1966 and 1987 (Figure 8.1). With increased chemical use comes an increased potential for pollution of off-farm surface water and groundwater. Furthermore, loss of remaining wildlife habitat continued: the U.S. Department of the Interior estimates that conversion to agricultural use accounted for 87 percent of U.S. wetlands lost from the mid-1950s to the mid-1970s (Tiner 1984).

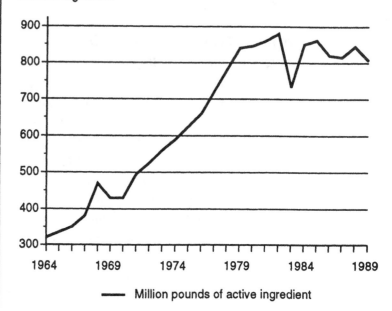

Figure 8.1. **Quantity of pesticides used in U.S. agriculture, 1964 to 1989.**

Million pounds
of active ingredient

— Million pounds of active ingredient

*Note:* Since 1982 the quantity has varied with fluctuations in the acreage under production. For example, pesticide use declined in 1983, largely because 78 million acres were taken out of production.

**Source:** Espelin, Grube, and Kibler, EPA, July 1991.

The nonfarm public has become alert to both issues, chemical use and wildlife habitat. Rising incomes lead urban citizens to put a high value on scarce wildlife habitat, endangered species, and the visual amenities of rural America. These changing attitudes have been reflected in public policy. Katherine Reichelderfer and Maureen Kuwano Hinkle have traced the evolution of government regulation of chemicals, from protecting farmers against fraud to more broadly protecting the environment and consumers.

Unlike the earlier proposals, which focused on contributing to the productive resources owned by farmers, the new proposals have implied threats. So the newer environmental issues of chemical pollution and wildlife habitat destruction contain the potential for more conflict between farmers and nonfarmers. Attempts to slow down this kind of environmental degradation directly raise farm production costs, with few benefits to farmers in return.

### Government Policies That Sometimes Encourage Environmental Degradation

Farm-price and income-support programs have actually contributed to environmental degradation. For example, commodity program benefits are tied to production of certain crops. This program design encourages farmers to grow particular crops and to produce more on every acre enrolled in the program. In addition to farm income and price supports, federal policies also provide rangeland and water resources to farmers at subsidized rates. These policies encourage farmers to use more chemicals, to cultivate or graze fragile land, and to exploit aquifers; they also discourage crop rotations (Table 8.1). In effect, farmers are paid to maintain production on fragile lands, to use scarce water resources, and to grow more per acre.

Of course, environmental degradation was not the aim of the offending policies, but an unintended and thoughtless consequence. As Katherine Reichelderfer points out, however, once these inconsistencies are recognized, and as public funds are increasingly limited, policy consistency becomes more important. Furthermore, as environmental goals become broader over time, old conservation policies become increasingly outmoded.

Table 8.1.
**Independent Short-Run Effects of Agricultural Policy on Environmental Quality**

| Agricultural Policy Instruments That— | Net Effect on— | | | |
|---|---|---|---|---|
| | Total Soil Erosion | Loss of Wildlife Habitat | Rates of Agrichemical Use | Total Use of Agrichemicals |
| Raise commodity prices | ↑ | ↑ | ↑ | ↑ |
| Tie farm income support to production levels | ↑ | ↑ | ↑ | ↑ |
| Reduce risk | ↑ | ↑ | ↓ | ↑ |
| Subsidize credit | ↑↓ | ↑ | ↑ | ↑ |
| Require short-term acreage retirement | ↓ | No effect | ↑ | ↓ |
| Establish cosmetic standards | No effect | No effect | ↑ | ↑ |

*Note:* Arrows indicate direction of net effect (increase or decrease) and do not imply whether the effect is "good" or "bad."

**Source:** Reprinted from Reichelderfer, 1990.

# The Renewed Importance of Stewardship

Of all the values discussed in this book, the most threatened is the primacy of farmers as stewards. Therefore, the most vociferously defended myth is the pastoral ideal, which has long been honored in American culture. It conjures up ideas and meanings that reinforce a society's image of itself, its inherent dignity, its basic goodness, and provides an ethical basis for evaluating its choices about social and economic development. Our attraction to pastoral symbols can be seen in the continuing popularity of great American landscape artists such as Frederic Church (Figure 8.2) or Thomas Hart Benton (Figure 8.3). Their images reach deep into the nation's collective psyche, what we believe to be good and right about our past. That ideal is therefore hard to let go.

The images of farm life as harmonious with nature are promoted in the books we read to our children. We teach them the animal sounds as Farmer Brown makes his way from pen to pen, and we teach them that the seasons are tied to planting, growth, harvest, and regeneration. These images also teach the virtues of independence, hard work, family, and community and that the natural environment is interwoven with those virtues. These values are certainly worth teaching, but they create mischief for the policy process when they are identified exclusively with the relatively few Americans who are still family farmers.

The fact that farmers do have economic and personal incentives for preserving our soil, water, and other resources has been essential in promoting sustainable agriculture. But that is not enough. We also need a better assessment of how national policies associated with modernization and commercial farming affect environmental quality. The public expects farmers to be all things: to be profitable, to be stewards of the environment, and to be producers of a cheap, safe food supply. Farmers, to their credit, have been willing to take on, if not always fulfill, these expectations.

That we need to formulate a workable set of expectations can be seen in the following paradox, one that is central to policy discussions. Two seemingly polar views of the environment exist. The first suggests that farmers cannot practice conservation in a political economy dominated both by large corporations and by the need to maximize profits. The second view hints that technological solutions such as chemical tillage, or no-till conservation using chemical cultivation, will take care of environmental problems. Both views hold enough common sense to be influential in the environmental debates. Yet each is flawed, primarily because of the power attached to the particular myth supporting it.

Figure 8.2. *West Rock, New Haven,* by Frederic Edwin Church.

Figure 8.3. *White Calf,* by Thomas Hart Benton.

## What Is Sustainable Agriculture?

The term *sustainable agriculture* has as many definitions as it has proponents. Common to all definitions is the basic idea that agricultural resources should not be used up faster than they can be replaced. Thus soil erosion should not exceed the capacity for soil renewal. Irrigation should not empty an aquifer faster than the water can be replaced by rainfall. The notion of sustainability also extends to the environment beyond the farm. For example, any chemicals used should not damage the natural ecosystem or resources such as groundwater used by other members of society. How to achieve the goal of sustaining agricultural resources is the source of contention.

Some adherents equate sustainable agricultural practices with organic farming, others with current farm practices. Perhaps the most widely accepted definition of sustainable agriculture is simply practices that minimize the use of chemical inputs and the loss of soil and water resources. Such practices include:

- Integrated pest management to minimize pesticide use
- Crop rotations to control pests and restore soil fertility
- Conservation tillage practices to reduce soil erosion
- Health management of livestock to reduce the need for antibiotics

As this list shows, *sustainability* is often defined in relative terms—that is, as reducing current levels of inputs that are alleged to be unnecessary and unsustainable. Because the extent of off-farm damage to the environment from agriculture is not fully understood (see Box 8.1), the sustainability of any particular technique is hard to establish. Thus a practical definition of sustainable agriculture is hard to pin down. While the goal of sustainability is widely accepted as worthwhile, how to reach that goal is likely to be a contentious issue for some time to come.

Certainly, farmers need to pursue profits. But the trick, in either a public policy or market solution, is to have farm prices reflect the costs of environmental degradation. Current farm programs confuse price signals and make some kinds of degradation more profitable. Further, the market does not reward farmers who are good stewards with higher prices.

Improving environmental quality and conserving resources in rural America may involve new technologies, but they are not absolutely necessary for solving our problems. Techniques that come closer to maintaining sustainable agricultural systems already exist (see Box 8.2).

Ironically, many of these techniques would be more profitable to adopt without current farm programs. We need to recognize the many factors causing environmental degradation. The solutions will require compromise and cooperation among various competing social interests. Public policy, as opposed to market solutions or voluntary adoption, is needed to bring the best possible chance for a sustainable U.S. agriculture.

Farmers continue to show their willingness to support conservation efforts. In fact, in areas where agricultural pollution of groundwater has been confirmed, farmers and their representatives have taken the lead in solving these problems. At the grass-roots level, farmers have been willing to participate, even though many of their national organizations remain contentious or just plain disinterested. This willingness needs to be tapped effectively.

Farmers must ultimately take responsibility, however, for the roadblocks that some farm organizations have placed in the path of environmental progress. Farm organizations have sought to maintain the status quo for commercial agriculture by defending what is actually in the interest of chemical companies. They have used the scare tactic of higher food prices to avoid changing current production practices. Farm groups must recognize that such tactics only alienate what appears to be an otherwise supportive general public.

Environmental groups, often long-established organizations but quite recent participants in agricultural policy, are also interested in sustainability. These groups bring to political debates a large nonfarm constituency on agricultural resource policy. Moreover, their financial supporters are not just from rural congressional districts and states, as are most farmers. During the 1985 farm bill negotiations, a period when farmers were going through a grave financial crisis, environmentalists mobilized considerable influence in return for support of expanded farm financial aid. In 1990, in the face of a farm backlash against environmental rules, these groups won several more demands, demonstrating their growing influence.

To maintain their influence, environmentalists will have to resist the temptation to foster a simplistic image of the family farm. Many activists tend to glorify the small-scale producer while unfairly attacking large-scale farmers. Besides being wrong, this attitude damages the potential for policy cooperation. Furthermore, these groups often fail to recognize the inherent conflict between income-support policies that encourage production and policies that discourage environmental degradation. This blinkered view may lead them to be co-opted by agricultural, as well as by other interests.

The continued influence of environmental interest groups will depend upon their ability to make informed decisions. But it will also depend on increasing their stake in the outcome of farm policy.

Environmentalists need to take partial ownership, and thus partial responsibility, for program success or failure. Interest group politics tends to favor those groups that not only have a knowledgeably expressed and vested interest in the programs, but also can deliver money and votes. Hence, as William P. Browne's research shows, farm and commodity groups have a much stronger position in the negotiating process than do environmental groups.

Correcting this imbalance demands greater involvement. To gain greater influence, environmental lobbyists must do the following:

1. Learn specifically how farm programs work
2. Develop alternative or complementary programs where possible
3. Replace values with facts as the substance of their lobbying
4. Understand farming for what it is today without any misinterpretations of myth
5. Decide on a reasonable and attainable political agenda that farmers can buy into and become a part of

Environmentalists must also cultivate supporters at the grass roots who are equally knowledgeable about agricultural problems, care about farm problems in the same way that bigger issues such as the Love Canal command attention, and, in response, generate political demands from the home district or state.

In short, agricultural producers and environmental activists share common interests. Even where these interests diverge, we can appeal to broader notions of the public good. True stewardship requires the cultivation of political goodwill, just as surely as it requires good farming.

# 9 Never Assume That Farm Programs Are Food Programs

An abundant food supply is part of our national self-image. Norman Rockwell's painting of the Thanksgiving dinner glows with a sense of pride in the nation's ability to produce food (Figure 9.1). Thanksgiving itself is symbolized by a cornucopia overflowing with foodstuffs. Feelings run deep, with human survival our central concern. "Freedom from want," in fact, sums up the feelings of earlier immigrant generations who found abundant natural resources in the New World. Productive farmers who produce an ample food supply at low cost are part of our American dream and are central to the development of the agrarian myth.

Although signs of agricultural abundance are also valued to some extent because they support belief in God's grace, it is more to the point for us to recall that many of the European immigrants who populated the American Midwest were fleeing famine and chronic hunger. These people came to the United States bent on securing stable food supplies above all else—and they went forward with a vengeance.

During the 19th century, public policy focused on distribution of land. Then, as the nation industrialized, urban workers began to suffer from food shortages. One of the most poignant chapters from John Steinbeck's *The Grapes of Wrath* depicts growers destroying food that was badly needed by the poor. Steinbeck describes the practice of destroying food that could not be sold at or above production costs as "a crime against nature." Modern American farm price policy was born in the Great Depression, an era when the chief problem was to ensure that food was available while balancing the income needs of farmers.

During the last half century, the easiest way to protect farm programs has been to confuse them in the public's eye with food policy.

Figure 9.1. *Freedom from Want,* by Norman Rockwell.

Because many people believe that U.S. agricultural policy contributes to America's abundance, contrived fears are readily sold in the marketplace of political demands. Some farm and rural interests, especially the ones identified with agrarian protest movements, have worked hard to portray farm programs as "cheap food" programs. As farmers have diminished in number and influence, both agribusiness and mainstream farm groups have increasingly sought political support for farm programs by appealing to consumer interests. With the exception of 1970, the theme of adequate food has continually appeared since 1965 in the titles of farm bills:

- *Food* and Agriculture Act of 1965
- Agricultural Act of 1970
- Agriculture and *Consumer Protection* Act of 1973
- *Food* and Agriculture Act of 1977
- Agriculture and *Food* Act of 1981
- *Food Security* Act of 1985
- *Food*, Agriculture, Trade, and Conservation Act of 1990

A prevailing argument, used by liberals and conservatives alike, is that family farms must be maintained so that large corporations will not control the food supply. "Big business"—presumably a General Motors of Food—would gouge consumers with higher prices and would encourage factory farming practices that produce unwholesome food. A large number of prosperous but competitive family farmers, the story goes, are necessary to ensure that wholesome food is available at reasonable prices. This argument is often stretched to imply that a decline in farm numbers will eventually translate into a lack of food. Thus consumers' alleged interest in maintaining family farms has been used to justify farm price and income support.

Related to anticorporate complaints is the argued need to maintain stable consumer prices. In this view, keeping resources in agriculture and encouraging surplus production is justified as insurance against years of crop failure, as in 1988, or surges in world demand, as in 1973. By keeping productive resources in place and maintaining government stocks, we have a cushion in case of a drastic drop in supply. Although no farm programs are designed specifically to limit rising prices, it is argued that the mere existence of surplus stocks puts a lid on food price increases and consumer suffering. Thus programs that encourage farmers to produce more than will be eaten serve as a brake on food prices.

U.S. policy also includes programs for feeding the hungry. Once it became clear that farm income would not be challenged by food giveaways and concessional sales, farmers began supporting programs "for the needy." These programs have indeed offered strategic policy advantages to farmers. In the 1960s, urban and farm congressmen joined forces to push for an expansion of domestic and international food relief programs. These programs were seen as a way to relieve hunger and boost demand for food and hence farm products. Some of the programs benefit the farm community by getting rid of specific crop surpluses, and they help the poor who need food. Because of their dual purpose, these programs have been accepted even by traditional rural conservatives who, on moral grounds, oppose social welfare programs.

## Why Farm Programs Are Not Food Programs

The apparent common sense in equating farm programs with food programs is misplaced. In fact, as time has gone by, the goals of farm programs and the interests of consumers have moved farther apart. We note four points where these two diverge:

- Farm production is only a small part of food production.
- Farm programs are not cheap food programs.
- Farm programs are not food price-stabilization programs.
- Food assistance programs are tied to farm program goals.

We need to elaborate on each point to explain why food programs must be determined apart from farm programs. But the essence of our argument is simple—the United States does not have a well-directed food policy. Rather, we have a collection of fragmented and often contradictory food and farm programs.

### Farm Production Is Not the Same as Food Production

The distance between farm production and food consumption grows greater every year. At the turn of the century, farmers grew wheat, millers ground it, and housewives baked bread from it. Farmers still grow wheat and millers grind it, but the wheat may end up in cookies, frozen pie crust, breakfast cereals, or a hamburger bun retailed at a fast-food outlet. Urbanization and rising incomes have gradually shifted food processing from the home to the factory. The result is that the farmers' share of the consumer dollar spent at the supermarket fell from 51 cents in 1918 to 24 cents by 1990.

In addition to 24 cents for farm products, the consumer's dollar buys another 76 cents of services, including transportation, processing, retailing, and restaurant service; the cost of labor takes the biggest single bite (Figure 9.2). Food processing and marketing employ more people and generate a much larger share of the GNP than farm production does (Table 9.1).

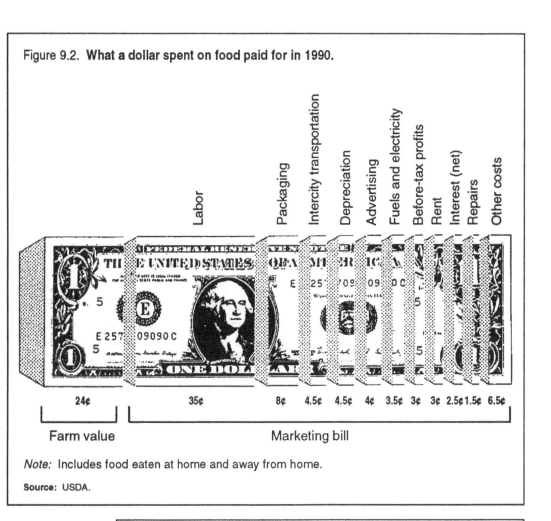

Figure 9.2. **What a dollar spent on food paid for in 1990.**

Labor · Packaging · Intercity transportation · Depreciation · Advertising · Fuels and electricity · Before-tax profits · Rent · Interest (net) · Repairs · Other costs

24¢    35¢    8¢   4.5¢   4.5¢   4¢   3.5¢   3¢   3¢   2.5¢   1.5¢   6.5¢

Farm value           Marketing bill

*Note:* Includes food eaten at home and away from home.

**Source:** USDA.

Table 9.1.
**Employment and GNP in the Farm and Food Sector, 1988**

|  | Farm Production | Processing and Marketing |
| --- | --- | --- |
| Employment, million | 1.8 | 16.4 |
| Share of GNP, % | 2.0 | 13.0 |

Because food marketing is a big part of the consumer's food dollar, agricultural abundance is only weakly reflected in the consumer's cost of food. Farm prices are just one of many factors influencing food prices. Changes in labor or energy costs can affect food costs even more than changes in farm prices can. A look at the past thirty years shows that food prices increased rapidly with the farm price boom of the 1970s (Figure 9.3). In the 1980s, food prices followed increases in marketing costs as consumers demanded more services (see Box 9.1), but farm prices stayed roughly constant.

### Farm Programs Are Not Cheap Food Programs

From the consumer's point of view, farm programs can be divided into three types: supply control, marketing orders, and price and income support. The third type takes up the most tax money, but it also has little effect on the prices we pay as consumers.

To regulate supply and marketing, supply-control programs raise the price of a commodity through import restrictions, government intervention, or both. This kind of program substantially raises consumer prices for only a small number of commodities such as sugar, dairy products, and peanuts. According to USDA estimates, these programs raised the U.S. consumer price of sugar by 43 percent and powdered milk by 47 percent (Webb et al. 1990). Between 1982 and 1986, consumers paid $6 billion for sugar and dairy policies. Together these commodities account for 10 percent of consumer food expenditures. Thus this type of farm program has a small but noticeable effect in raising the cost of our food.

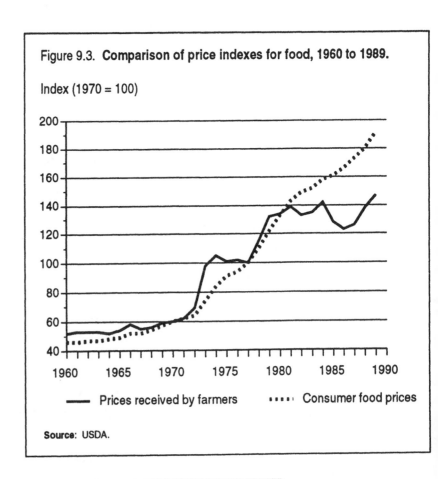

Figure 9.3. **Comparison of price indexes for food, 1960 to 1989.**

Index (1970 = 100)

Legend: —— Prices received by farmers     ····· Consumer food prices

Source: USDA.

## Are "Large" Marketing Margins Bad for Farmers?

Some see the trend towards larger food industry margins as bad and blame this trend for depressed farm incomes. This is misguided. The farmer's share of the dollar does not indicate either the size of gross income or net profits at the farm level. In 1988, the farmers' share of the consumer dollar set a new record low, while net farm income reached a record high.

The costs of marketing margins arise from the desire of U.S. consumers to purchase more services along with food. If consumers were to purchase everything directly from farmers, it would not raise farm income, but would only reduce food industry output. Given the rising numbers of single-person households, single-parent households, and women working outside the home, it seems inevitable that American consumers will continue to seek more food marketing services of all kinds. Demand for these services will grow much more than the underlying demand for farm products. Thus food marketing margins will continue to grow.

The second kind of farm program is the federal authorization of marketing orders for fruits and vegetables. These orders allow producers of perishable crops to organize in order to control the quantities marketed or produced. The effect of these programs has been controversial. Proponents of the programs point to their stabilization features, while opponents argue that they are used to raise consumer prices. It is clear, however, that fruit and vegetable marketing orders have not reduced consumer costs for fresh produce.

The third type of farm program is the set of price- and income-support programs for crops, including wheat, feed grains, and rice. Receiving the bulk of government expenditures, these programs constitute the heart of U.S. farm programs. Although the prices of these commodities are a small share of retail food costs, they do influence consumer prices of meat and poultry, which account for 18 percent of consumer expenditures on food.

Sorting out the effects that these programs have on food and feed grain prices is not simple. On the one hand, programs should tend to reduce prices by encouraging production through direct payments to producers. On the other hand, programs should tend to increase prices in that the support price becomes the minimum price that any producer receives. Rodney Tyers and Kym Anderson, estimating the effects of removing U.S. farm programs in the 1980s, found that market prices would fall very slightly for feed grains and change little for wheat and rice. Price- and income-support programs for grains have probably caused only slight changes in the consumer cost of food.

### Farm Programs Are Not Food Price-Stabilization Programs

If farm programs have not been designed to reduce food costs, have they at least provided us with stable food prices? As James Houck notes, stability of food prices is a universally desired goal, but it is difficult to accomplish. One reason is that farmers and consumers have different interests in price stability. Farmers wish to be protected from sharply falling prices, and consumers wish to be protected from sharply rising prices.

Farm program rules are designed to insulate producers against falling prices. This insurance for producers is provided by the Commodity Credit Corporation (CCC), which acts as a buyer of last resort at support prices. CCC takes possession of commodities when market prices are lower than the support price and is authorized to sell them domestically when market prices reach 150 percent of the support price (Figure 9.4).

When consumer prices have risen rapidly, as in the early 1970s, there has been public pressure for price containment. Because CCC stocks can be quickly exhausted, other means of intervention are needed to contain prices. For example, the U.S. government suspended the import quota on beef and placed an embargo on soybean exports in 1973 in order to reduce domestic food prices (see Box 9.2). Although farm groups frequently argue the motto "stabilization," both of these actions brought strong protests from producers because they reduced profits.

> *Government programs have also worked against price stability when sudden changes in policy coincided with unexpected market events.*

Government programs have also worked against price stability when sudden changes in policy coincided with unexpected market events. The payment-in-kind (PIK) program of 1983 is a recent example of poor timing. In an attempt to reduce surpluses, the government paid farmers in kind from surplus stocks for not planting. A severe drought also reduced production, and the result was a temporary sharp price increase.

Price stabilization has much intuitive appeal, but it is a very difficult policy goal. Although neutral stabilization policy would contain prices around the long-run average market price, that average shifts over time. On the one hand, if stabilization policy is controlled politically by farm interests, prices are likely to be stabilized above the market average. When this happens, the government must subsidize the disposal of surplus stocks (Figure 9.5). On the other hand, protecting

Box 9.2.

## Management of CCC Stocks

A look at the past performance of government stock operations is enlighten-ing. The loan rate, market price, and CCC minimum release price for corn are shown in Figure 9.4. The loan rate and release price essentially set a band within which market prices should fluctuate. The unusual surge in world demand in the early 1970s caused commodity prices to rise sharply. CCC stocks were quickly exhausted, and prices could not be contained. Loan rate and release prices were then adjusted to the new level of market prices, and the release price was set at a larger percentage of the loan rate. From 1976 to 1983, market prices stayed within the band, but were most often at the loan rate floor. In the mid-1980s, prices fell below the floor as surplus stocks were released through PIK (payment-in-kind) certificates issued as payments under farm programs.

Market prices have been at or below the loan rate quite often during the last fif-teen years. By defending support prices above the long-run market price, the U.S. government accumulates surplus stocks and then must find a means to dispose of them. Surpluses have been disposed of by donating stocks to domestic and for-eign food relief programs, subsidizing export sales, or using stocks to make direct payments to farmers (Figure 9.5). Release of stocks for PIK payments pushes mar-ket prices down and works against stabilization goals. Even so, PIK was used to reduce the very high level of stocks that had accumulated in the mid-1980s.

CCC's record shows the difficulty of pursuing price stabilization. Big increases in demand, as in 1973, can quickly exhaust stocks. Slumps in demand, as in the early 1980s, can create a surplus that is expensive to store or dispose of. Major swings in market demand can quickly exhaust CCC's ability to either release or hold stocks.

consumers against disastrous price increases is also costly. The food price escalation of 1972-74, while painful for American consumers, was highly unusual. Holding enough stocks as insurance against price increases that happen once in thirty to fifty years would be so expensive that it would be impractical.

The bottom line is that farm programs are not intended to ensure cheap food at stable prices, nor do they do so. Government inter-vention has sometimes *raised* farm prices through acting as buyer of last resort, authorizing supply control or price discrimination, or limiting imports of some foods. When prices have risen rapidly, these intervention tools have been used on behalf of the consumer, but with protests from some farm groups. Farm programs are

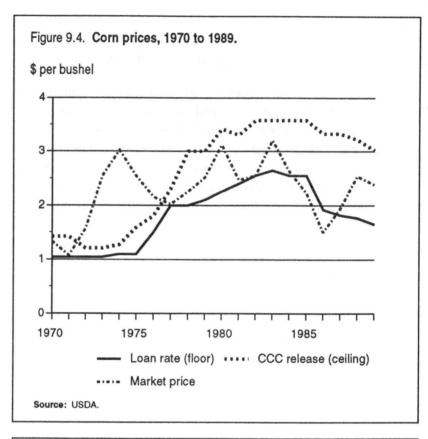

Figure 9.4. **Corn prices, 1970 to 1989.**

$ per bushel

Loan rate (floor) ····· CCC release (ceiling)
····· Market price

Source: USDA.

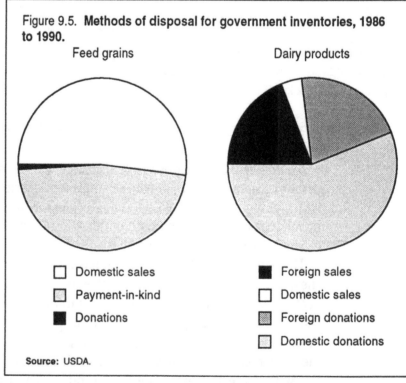

Figure 9.5. **Methods of disposal for government inventories, 1986 to 1990.**

Feed grains                    Dairy products

☐ Domestic sales          ■ Foreign sales
▨ Payment-in-kind          ☐ Domestic sales
■ Donations                ▨ Foreign donations
                           ☐ Domestic donations

Source: USDA.

designed primarily to support farm prices and incomes—not to limit increases in consumer prices.

### Food Assistance Programs Are Tied to Farm Programs

People tend to accept farm policy as food policy because they assume that the poor will not be deprived of food simply to protect farmers' incomes. Farmers supported this value judgment once they were assured of a correlative assumption, that the poor would not be fed solely at the farmers' expense. These twin assumptions led the way to tandem policies that assured farm income, while assuring food availability to the poor with other means. Although food consumers, including the poor, have not had their interests in cheap or stable food prices reflected in farm income policies, the opposite does not hold. Instead, domestic food assistance programs in the United States have always been politically linked to agricultural programs. Food assistance programs have combined the goals of relieving hunger, boosting demand for farm products, and reducing the government's cost of holding stocks. These programs fall into two general categories: commodity distribution and food stamps.

Operating since the 1930s, commodity distribution programs became increasingly important during the 1980s. The amount distributed is determined by the availability of publicly held food stocks that accumulate as a result of federal price-support programs for some farm products. The amount is not determined by need. Although the recipients may be identified by a means test, the amount distributed is determined by surplus stocks rather than by the number of hungry people. These distributions support the goals of farm programs by reducing the cost of holding public stocks. Because surplus stocks were larger in the 1980s, distributions went up sharply.[1] Government costs for these programs were $5 billion to $7 billion annually during the 1980s, including the value of surplus commodities.

During the 1960s, a major new food program was introduced that did not rely on surplus disposal. The food stamp program, which had operated briefly during the 1930s, was resurrected as part of the 1960s War on Poverty. The program was permanently funded in 1964, and its nationwide operation was mandated in 1974. Unlike commodity distribution programs, the food stamp program is an entitlement program. Anyone who qualifies may receive benefits in the form of coupons that can be exchanged for food at grocery stores.[2] During the 1980s, food stamp benefits of about $11 billion were distributed each year to some 20 million people.

Food stamps do increase demand for food, but their impact is very limited. USDA has estimated that food stamps increase total farm commodity demand by only 1 to 2 percent. Their impact is limited by two factors. Food stamps do not usually increase household food purchases by the full amount of the coupon allotment. The coupon benefits allow the household to shift some money that would have been spent on food to nonfood items. In addition, as we discussed above, retail purchases return only about 24 cents out of every dollar to farmers.

The rationale for linking food and agriculture was to avoid repeating the crime of simultaneous food abundance and chronic hunger. The link was a political compromise that protected farm interests. We have shielded U.S. farmers from the impacts of greater food supplies and lower food prices, while U.S. consumers have benefited (see Box 9.3). U.S. hunger relief programs are at best only mildly positive in their effect on farmers.

The key question, however, is whether these programs have adequately met the goal of assuring that food is available to the poor. We have reason to think that in the last decade protection from the ravages of hunger eroded in the United States. The most visible sign is the increasing number of homeless people. Being rootless, this new underclass has had fewer traditional welfare benefits such as food stamps. Only recently has policy returned to food needs, but not with much fervor.

First, the 1990 budget reconciliation cut farm programs substantially, but left food programs intact. With total expenditures of nearly $20 billion annually, food programs will become more important in USDA's budget than farm programs, which now spend $10 billion to $15 billion. For the two previous decades, these expenditures were, by political agreement, roughly balanced. Another, less important change was the 1988 decision to fund commodity distributions under the Emergency Food Assistance Program through specific purchases. Authorizing funds for commodity purchases represents a significant departure from the past policy of allowing distributions to be determined by stocks. Finally, since 1988, households receiving Aid to Families with Dependent Children (AFDC) or Supplemental Security Income (SSI) are automatically eligible for food stamps.[3]

## Fragmented U.S. Food Policy

One thing should be clear so far. Farm programs are not designed to be food programs because their primary goal is to support farm incomes. We do have food programs, however. As policy, they consist of a fragmented and uncoordinated set of governmental efforts to address the concerns of consumers. We can put them into four categories:

- Food production technology policy

Box 9.3.

## Does America Have Cheap Food or a Cheap Food Policy?

American consumers spend a smaller portion of their income on food than do consumers in most other countries (Table 9.2). But does this mean that we have cheap food? The portion of income spent on food only indicates that U.S. food is cheap *relative* to our per capita income, which is one of the highest in the world. As income goes up, consumers spend a smaller *share* on food. This relationship between income and food expenditures is also true for households within this country: households with incomes below $10,000 spend more than 20 percent on food, while those with incomes above $50,000 spend less than 10 percent. On average, U.S. consumers are well-off compared with consumers in other countries because our high incomes allow us to spend more on luxuries and less on necessities like food.

Is the price of U.S. food cheaper than in other countries? The price of *some* foods is lower here than elsewhere (Figure 9.6), and the United States is the world's least-cost producer of many agricultural commodities. Two elements of agricultural policy have contributed to these lower prices. First, we have been generously endowed with agricultural land, and federally funded agricultural research has increased its productivity, thus lowering food costs over the long run. It now takes fewer resources to produce farm products. This is part of why national income has grown and why less of it goes for meeting basic food needs. Funding for agricultural research has been a major thrust of farm policy since the Civil War, and it continues to account for nearly $1 billion of federal spending every year.

Second, most farm policies in the United States do not support farmers by forcing consumers to pay higher prices for food. Most government transfers to farmers are direct payments. With a few exceptions, farm incomes have not been supported through higher consumer prices. Direct payments come about because producers of wheat, rice, and feed grains are paid the difference between the market price and a target price. These direct payments accounted for one quarter to one half of government expenditures on farm programs during the late 1980s.

Thus the design of most farm programs has placed the burden of program cost on taxpayers, rather than on consumers. Because the poor spend a larger portion of their income on food and pay fewer taxes, the burden of farm program costs falls on the richer members of society. The design of U.S. policies is in sharp contrast to policies in Europe and Japan, where consumers pay higher prices for food in order to support farmers. But does the United States have a "cheap" food policy just because consumers are not forced to pay higher prices?

Table 9.2.
**Food Purchases as Percentage of Income in Selected Countries**

| Country | Per Capita Income (1986 $) | % of Household Income Spent on Food (1980-85) |
|---|---|---|
| **Low income** | | |
| Ethiopia | 120 | 32 |
| Mali | 180 | 57 |
| Tanzania | 250 | 62 |
| India | 290 | 52 |
| **Middle income** | | |
| Indonesia | 490 | 48 |
| Peru | 1,090 | 35 |
| Tunisia | 1,140 | 42 |
| Colombia | 1,230 | 29 |
| **Upper-middle income** | | |
| Poland | 2,070 | 29 |
| Argentina | 2,350 | 35 |
| Israel | 6,210 | 26 |
| Singapore | 7,410 | 19 |
| **Industrialized** | | |
| France | 10,720 | 16 |
| Germany | 12,080 | 12 |
| Japan | 12,840 | 19 |
| United States | 17,480 | 13 |

**Source:** World Bank.

- Antitrust legislation
- Food relief programs
- Food safety and nutrition policies

These four areas of policy do indeed focus upon consumer needs. But because they can also influence farm income, farm interests are often vocal in debating their direction.

### Support for Cost-Reducing Technological Change

Support for agricultural research is an important part of agriculture policy, and this support was expanded slightly in 1990. Improvements in productivity do benefit consumers by reducing food costs (see Box 9.3). They also benefit U.S. agriculture by making exports more com-

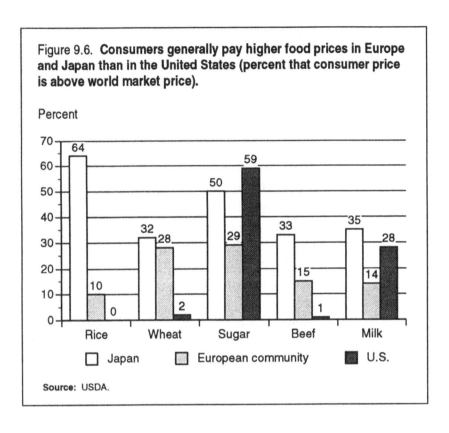

Figure 9.6. **Consumers generally pay higher food prices in Europe and Japan than in the United States (percent that consumer price is above world market price).**

Percent

Source: USDA.

petitive. Because food costs now take a smaller part of consumers' budgets, the emphasis of this policy should shift in the future. Research to improve the safety, quality, and nutritional value of the food supply could have higher returns.

## Enforcement of Antitrust Laws in the Food Industry

The big margin between farm and consumer prices often leads observers to question whether food industry profits are too large (see Box 9.1). While most of that margin can be attributed to marketing costs rather than to profits, rising concentration of market power in the food industry is a cause for public concern. The food industry has certainly not been exempt from antitrust laws. Infant formula and beef are two products that have come under recent scrutiny by the Department of Justice.

## Food Relief Programs for the Needy

Although the United States has a relatively abundant food supply, many families and individuals cannot afford an adequate diet. More than 30 million Americans remain below the poverty level. Food programs are part of the safety net for these people. Debate about the design and delivery of commodity distribution and food stamp programs obviously continues, but we believe that distributions should be determined by need rather than by the availability of surplus stocks.

### Efforts to Ensure Food Safety and Nutrition

The growing consumer concern with food safety reflects the rational perceptions of a well-off and aging society. Now that food quantity is no longer a concern for many of us, food quality has been highlighted. New scientific evidence increasingly points to a link between diet and the development of many diseases. As people live longer, these links become more apparent and more important. Thus Americans are increasingly concerned about food safety and nutrition and are demanding greater public intervention to regulate hazards and promote nutritional education.

Food safety and nutrition policies stir up both comment and opposition from producers, who have objected to unanticipated regulatory actions such as the banning of Alar and who have contested dietary guidelines that might have encouraged Americans to eat less meat and other sources of fat. The 1991 controversy about USDA's proposed revisions of a nutrition chart, which downplays the role of livestock products in the diet, underscores the conflict between consumer and industry interests. Issues such as these emphasize contradictions inherent in the belief that farm and food programs must be merged to serve common interests. The history of these programs suggests they cannot. Hard choices must be made on behalf of our society's food needs.

## Compassion and Fairness

Linking consumer interests to farm programs has increased the support for both farm and food assistance programs, but the principal winners have been farm program beneficiaries, who have gained broad-based support for farm legislation. At the same time, they have also controlled the consumer and food assistance agenda legislated through the farm bill and administered through USDA. The poor, as consumers, have benefited in general from food assistance programs, which have boosted the overall level of government benefits they receive. But the poor have lost to the extent that their choices have been limited by the design of food programs or when benefits have fluctuated with stock levels. Other losers are the taxpayers, who have footed the bill for farm income and price support, and consumers, who have paid higher prices for some food items.

The interests of farmers and consumers no longer converge as easily as they once seemed to. Farming and food are not synonymous. Farm income, as explained in Chapter 7, is increasingly determined by export demand rather than by domestic demand. Consumer food prices, in contrast, are determined increasingly by the costs of marketing and processing. Furthermore, over time, rising incomes have meant that a smaller percentage of our income is spent on food. Controlling food costs is less vital for us than it once was. Diet and health concerns have become more prominent. Thus the links between farm programs, farm prices, consumer prices, and the food needs of the poor—or between consumer demand and farm prices— are less strong than they used to be.

> Compassion for the hungry must be accompanied by fairness to food producers.

Nevertheless, we must not lose sight of the wisdom expressed in New Deal responses to the "crime" noted by Steinbeck. New Deal farm policy was intended to assure that market forces did not keep food out of the mouths of hungry people, while also assuring that people were not fed at the expense of food producers alone. The values that serve as our rationale for linking farm and food programs might be expressed in terms of *compassion* and *fairness*. Compassion for the hungry must be accompanied by fairness to food producers. This principle is still as viable as it ever was and is still as crucial to the evaluation of food policies. Furthermore, it is reasonable to include producers in the process for establishing food policies. What better way than participation is there to assure that farm interests will be dealt with fairly? But should this mean that we must remain committed to the existing instruments of farm policy, which only coincidentally serve food needs? Of course not. Given the evidence throughout this book that current policies are not serving either farm or food interests, we have ample reason to consider serious reforms.

## Notes

1. Distributions are made under the National School Lunch and Breakfast Programs, the Child Care Food Program, the Nutrition Program for the Elderly, the Temporary Emergency Food Assistance Program, the Commodity Supplemental Food Program, the Food Distribution Program on Indian Reservations, and through charitable institutions.

2. The value of the coupons is equal to the value of a low-cost diet (as determined by USDA) minus 30 percent of the household's "net" income (after rent and other essential expenses). The idea behind the coupon allotment is that poor households normally spend 30 percent of their income on food. Food stamps then provide the additional expenditures needed for minimum nutritional requirements.

3. The more fundamental question of how best to deliver services to the poor is beyond the scope of this book, but it is notable that changes in food stamp delivery are being tested. The Family Welfare Reform Act of 1987 and the Family Support Act of 1988 are testing new methods of delivering services to the poor, including "cashing out" food stamps. In a five-year demonstration project in the state of Washington, welfare payments and food stamp benefits are combined into one cash payment.

# 10 Never Assume That a Government Program Will Do What It Says

Even the most casual readers of earlier chapters probably noted two things of importance about U.S. agricultural policy. Both are recurring themes that anyone who deals with agricultural policy must understand in order to interpret policy outcomes effectively.

First, *the range of individual programs is incredibly broad, and the number of programs is enormous.* Agricultural policy is not just farm policy. Programs also are in place to promote food availability, fiber supplies, food and fiber industry growth, food safety, product regulation, nutrition, domestic food assistance, food assistance to the world's poor, international trade, rural development, welfare of rural residents, resource conservation, environmental protection, and animal welfare. We could go on.

Second, *the logic behind any one program is often flawed because there are so many purposes.* As programs have grown in number over the last several decades, the public rationale for expansion has been stretched to the limits of believability. Each addition, so the story goes, contributed to a sound national policy to serve the public's well-being. In reality, new programs arose from the political need to buy new constituents in support of old policy goals. This, more than any other factor, has perpetuated the myths. New programs were added primarily to keep farm programs alive. By and large, national policy goals took shape only after the deals were cut to justify what was done; the goals were not the plans behind the actions.

## The Myths of Public Policy

So far we have concentrated on agrarian myths, but we are influenced by myths about government too. Two of the most powerful are

(1) the belief that government can and should meet social needs through a carefully formulated plan of attack and (2) the belief that government cannot meet many needs and should not even try to do so. Ironically, each of these myths encompasses assumptions for political philosophies published by two Englishmen in 1789, about the time the U.S. Constitution was being ratified. Jeremy Bentham argued that law and policy should be based on an assessment of the likely consequences. Legislators should enact laws that produce optimally beneficial outcomes that help the greatest number possible while harming the least. Edmund Burke, on the other hand, argued that people could find the optimum without government help. If areas of public life appeared to fall below desired performance levels, the responsible experts had probably overlooked the actual consequences of legislation.

Because many Americans hold one or both of these beliefs, we have contradictory expectations for government. On the one hand, we expect government to proactively create and administer public policies to alleviate public problems. On the other hand, we expect things to go pretty well without government intervention and expect that regulatory policies will create a nuisance at best. Given the contradictory nature of these assumptions (see Figure 10.1), it is not surprising that we often get contradictory results.

*We have contradictory expectations for government.*

In agricultural policy, the *myth of optimal planning* is seen in the belief that farm-state legislators have a moral duty to bring home government programs that reduce financial risk, guarantee income, and preserve valued local institutions like schools, hospitals, and rural communities. The *myth of government ineffectiveness* is expressed in complaints about environmental regulations and paper work and in a vague but deeply held commitment to the free market. For many people outside the farm community, agricultural policy means rationally managed programs to assure food security and resource conservation. Or, alternatively, agriculture is a bastion of unregulated free enterprise, and the family farmer is a symbol of the virtues produced by an open economy. Elected officials, bureaucrats, and administrators are themselves often infected by these myths, seeking the advice of agricultural and policy experts, then rejecting that advice in favor of signals from a political pulse that represents the heartbeat of an America beyond the Washington beltway.

## A Short History of Confusion

Agricultural policy has generally been driven by faith in planning. The contrary faith in unplanned spontaneity has also had its day, however. The best-laid plans have been sent astray by shifting blocks of voters, outside shocks to the agricultural economy, and the election or appointment of individuals who are ideologically committed to limiting government programs (see Box 10.1). A short review of how we got where we are today will prove useful.

### The Evolution of Agricultural Programs: Stage One

When government first began to act on the basis of agricultural wants, the United States was a rural nation. A plan was not difficult to imagine. To satisfy the democratic majority, land was made available and an extensive agricultural establishment was created. The Homestead Act of 1862 legitimated private squatters on those public lands that investors had not claimed. But then, as before, westward expansion was fueled by yeoman farmers and ranchers who lacked both the insights and skills that make for successful producers. Accordingly, government stepped in to stabilize the expanding but often failing farm sector.

In 1862, the Department of Agriculture was created to oversee this policy. To educate the children of farmers, land-grant colleges for

---

Box 10.1.

## Contradictory Programs

Many government programs can be contradictory. For example, many agricultural programs have been focused on restricting acreage during times of surplus production. Meanwhile, other programs have contributed to surpluses, such as public research to increase productivity or federally subsidized water for agriculture in the West. In one sense, public programs are often analogous to putting one foot on the accelerator while the other is on the brake. At the same time that many programs have contributed to production and surpluses, other mechanisms like acreage reductions or set-asides are used to reduce production.

Another significant conflict in U.S. agricultural programs involves the policy on disaster assistance. During the 1970s, Congress passed free disaster assistance legislation that would provide federal dollars to farmers in areas that suffered from severe crop losses. In 1980, Congress passed the Federal Crop Insurance Act, making crop insurance the disaster assistance policy. In doing so, they explicitly stated that future disaster assistance would not be provided to farmers who farmed in areas where they could have purchased federal crop insurance. Since that time, Congress has passed "emergency" disaster assistance legislation no less than four times.

---

Figure 10.1.
**Bloom County,**
by Berke
Breathed.

1988, Washington
Post Writers Group.
Reprinted with
permission.

agriculture and technology were established the same year by federal bequests of property. As the states built schools based on these Morrill Act grants, supplements to education were judged useful. In 1887, the federal government again intervened and created experiment stations for state-of-the-art research on behalf of farm modernization. Later, in 1914, the Extension Service with its network of county agents was established to bring education and research to the isolated farm.

With such supports in place, the hope was that an indigenous farm population could develop on its own, thus avoiding the pattern of states such as Michigan, which had to actively recruit European farmers when its earliest pioneers faltered. Knowledge was seen as a key to avoiding poverty in agriculture. What characterized this first stage of policy development was its simplified approach to the public well-being. Assistance was made available to most people in the sector, as long as they wanted to avail themselves of its services. That establishment even made noticeable strides in providing for women and, with the creation of the 1890 land-grant schools, African Americans. By the end of the century, the government also began passing legislation that benefited consumers by regulating food safety.

### New Policy: Stage Two

The Great Depression, and 1933 in particular, marked a policy departure for U.S. agriculture, a useful one at the time, but one that became increasingly hard to justify as time passed and the circumstances changed. The Agricultural Adjustment Act (AAA) broke from the tradition of providing opportunities for farmers and their families. In the face of a depression that had slowed U.S. agricultural development since the early 1920s, government started a new public policy endeavor for six basic crops.

Quite simply, government began direct cash payments to producers, all in the name of keeping them in business so that food would be available. When prices fell below a certain level, farmers got money. AAA was an entitlement for farmers only because they were farmers and different from other segments of the population. For anyone who received the payment, other justifications aside, it was a supplemental income transfer. People like getting checks from the government because of their special status. So, not surprisingly, the number of program crops expanded as other producers clamored for a share. At the time, the preponderance of farm-state legislators in Congress made this easy. No plan was necessary.

### Burgeoning Policy: Stage Three

The farm depression of the 1920s and 1930s disappeared as World War II generated a high demand for food. Cash payments to farmers did not disappear, however. Of course, no plan was necessary to keep these payments alive when farmers were allowed to express their view on the need for federal dollars. When chronic oversupply returned in the late 1940s, the old rhetoric about keeping farmers in business was pulled out and repeated through the severe recession of the late 1950s and into the better times of the 1960s. Despite continually changing conditions, farmers were never again left on their own to fail in business.

Of course, other sectors of the economy also faced post-war failure from poor business practices. And, of course, political representatives of those other sectors wondered why the special people got special payments, especially since the electoral power of the declining farm population was shrinking, and absolutely no evidence indicated that all those payments helped keep farmers in business.

A new type of politics, described by Weldon V. Barton, emerged to keep farm payments alive in the 1960s. The era of the policy add-on brought farm bills and other farm legislation a series of new, if not better, reasons to support farmers. While trades were made with farm-state legislators to get their vote for such things as labor legislation, a more lasting, even institutionalized, strategy evolved for getting others to vote for farm policy.

The add-on strategy worked most obviously and was first apparent with food stamps. As we noted in Chapter 9, food stamps became a permanent part of the U.S. Department of Agriculture's job in 1964. With that program tacked on to it, USDA gained a new constituency that it could benefit from as well as serve. Farm-state legislators naturally gained support from liberal urban members of Congress who felt it essential to provide food benefits for the poor of their districts. The add-ons were so important that agricultural committee members of Congress, recalling how liberals had gushed earlier over the adoption of the Food for Peace Program in 1954, vowed their continued use. They also remembered, of course, that this program had been intended to dump available food supplies overseas while just incidentally making the United States look good in the world press.

The following decades spawned farm bills designed like theme parks, providing the same old ride with a new facade. Environmental protection, animal rights, and rural development initiatives pumped new thrills into farm bill politics and led to new programs, but the basic purpose of protecting direct cash payments to farmers was still the same old roller coaster.

*The following decades spawned farm bills designed like theme parks, providing the same old ride with a new facade.*

## Why Programs Can't Do What They Claim

The role of government has not always been so pervasive, as these three stages of government involvement in agriculture reveal. In the past when institutions such as the family and church made society more self-sufficient and the economy less interdependent, the United States operated more nearly as the so-called free enterprise system. Such a market-oriented system gave way from the very beginning to increased government intervention. Indeed, government has learned well that the market does fail. When failures occur, government intervention may become desirable either for economic or for political reasons, such as easing economic adjustments or transferring benefits from one group to another. Markets are not always kind. When they hurt one segment of society or another, the odds of government intervention increase. What then is the distinction between markets and government? When do markets fail? Why do we have government failure? What forces shape the role of government, and why has the role of government increased?

Without our being cynical, adoption of agricultural programs must be explained as politically inspired. Adoption is not driven by a well-focused needs assessment or even by the specific identification of market failure. Agricultural policy development is rational only to the extent that it follows an internal logic of salvaging as much as possible of the old order of political power in agriculture. The driving force is institutional defense. That's why policy lacks internal consistency, straightforward discussion of program goals, and conscientious forethought about the consequences of new programs. Therefore, the policymaking process continually adjusts at the margins—but only with frustrating slowness. Discussion of national agricultural needs may never take place. In fact, such discussions are often purposely avoided.

> *In order to understand government failure, we need to understand market failure.*

Policymakers, as we have emphasized throughout this book, should continually question the role of government in a U.S. political economy that is driven both by democratic wishes and the resulting protection of private and public sector enterprises. The basic question centers around the proper mix between markets and government. But in order to understand government failure, we need to understand market failure. The latter must be recognized when it exists, but it must not be blamed when other features of the political economy cause problems.

## Market Versus Government

The distinction between market and government is murky at best. A market system totally free from government involvement is a misnomer. Market systems require ownership of resources and the right to collect returns from those resources. So government has to be involved from the outset to establish some public agreement on protecting the ownership rights of individual entrepreneurs. The way these rights are established has definite repercussions for market performance (in other words, for the allocation of resources to their most valuable and best use) and for the distribution of wealth in a society.

Inheritance laws are a good example. By allowing wealth to be transferred from one generation to the next, they reflect the extent to which governments are willing to protect individual property rights. This type of intergenerational transfer of wealth leads to greater concentration of wealth in a society such as ours with a relatively low birth rate.

Because markets must depend upon governments to establish property rights and the rules for trade, how can markets and government be separate? When most people use the term *free market*, they are probably thinking of the lack of government intervention in influencing supply and demand and, thus, prices. But this assumption must be viewed cautiously. The rules for trade and the way property rights are protected can influence supply and demand. Thus the question is not one of markets or government. Rather, in an economy where citizens influence political decisions, the fundamental question is: What mix between markets and government can best reflect the preferences of the members of society? How do we establish property rights? Finding that mix is an ongoing process. Given our basic ideology of free enterprise, the burden of proof generally favors the market. This means that identifying market failure is essential if we are to understand when government action may be desirable.

## Market Failure

Prices drive a market system. Producers use price to judge what and how much to produce. Consumers use price to judge what and how much to consume. Consumer preferences, as long as they can be satisfied with buying power, affect price directly. Thus markets are said to satisfy society's preferences. When price operates as it should, economists have what they call efficiency. In the ideal, resources are put to their highest and best use, and no reallocation of them would improve the total output of society. In reality, however, a price-driven market system does not always correspond to what society wants at any given time.

When certain types of output from a production process do not enter the accounting system of a firm, and therefore have no corresponding price, the market has failed. Output of this sort can be thought of as by-products because the firm produces them unintentionally. Desirable or undesirable by-products can be produced. If they are desirable to everyone, government action is unlikely. If they are undesirable to someone, government action may be called for. For example, farmers spray their corn with pesticides, and these harmful chemicals enter a nearby stream. Clearly society does not want harmful chemicals in the waterways and, just as clearly, farmers do not want to pollute the stream. Yet the price that farmers receive for their corn includes no premium if they do not spray. On the contrary, farmers stand to lose income by forgoing spraying. The market fails to reflect society's preferences. But if the negative effects of chemicals touch the right people, the government becomes involved by regulating chemicals.

Noncompetitive markets are possibly the most pervasive example of market failure. The belief that markets can meet the preferences of society depends on competition. Simply defined, a competitive market is one with many buyers and many sellers. When only a few sellers (oligopoly) or only one seller (monopoly) is present, the seller can ensure a higher profit by restricting output to levels below what a competitive market would produce. Price is driven higher, and society then has fewer goods than it desires. In today's society, we have few examples of perfectly competitive markets. Accordingly, because grievances can always be easily expressed, the United States has a long history of government intervention in noncompetitive markets.

Market imperfections are a final category of market failure. When people cannot easily begin a new business or when resources cannot easily be moved from one production process to another (immobility through fixity of assets), market outcomes will be inefficient. In addition, if market information about production technology, market opportunities, and prices is not equally available to all market participants, society will produce less from its resource base than is technologically possible. In part, many of the tax laws that allow capital goods to be rapidly depreciated are designed to lower mobility barriers associated with technologically obsolete machinery. For the same reason, the government has also tried to relax the entry barriers for beginning farmers through such things as subsidized interest rates.

Although most economists do not include issues of distribution when discussing market failure, most acknowledge that the distributional consequences of a market economy may not correspond to society's preferences. There can be trade-offs between efficiency and equity. If income distribution is highly skewed, people may still starve, even though a society has allocated its resources efficiently. When policymakers choose a progressive tax system to redistribute wealth, they also run the risk of reducing the amount of resources devoted to production activities. Many of the original farm price programs were designed to redistribute wealth from taxpayers to farmers. In fact, as a review of the three stages of government involvement in agriculture reveals, most government action is just as concerned with who gets what and who pays as it is with how resources are organized to get the most output. Distributional issues obviously are a primary concern of politicians.

### Government Failure

Our discussion of market failure should not be misinterpreted to mean that government solutions are always superior. Government also fails. And it does so for a number of reasons. On occasion, market failure may be more consistent with society's preferences than is governmental action with its corresponding failure. Usually, examples of government failure can be cited only after the fact. It is extremely difficult to judge beforehand whether the government solution to market failure will be worse than the failure itself. Trade-offs are common, and distribution issues are nearly always involved. Government decides whose preferences count, which makes it difficult to identify government failure under any circumstance. Society's preferences are revealed by default through policy action in a democracy.

> *Our discussion of market failure should not be misinterpreted to mean that government solutions are always superior. Government also fails.*

To highlight how government fails, we present below six problems inherent in the governmental process in agriculture. Each of these problems helps us to understand why agricultural policy lacks the relevance that it might have in an ideal world. Each also explains why no one can ever know what a program actually does by listening to the rhetoric surrounding it.

*1. The Implementation Problem: Bureaucrats Have Their Own Incentives.* The legislature makes laws; the executive carries them out. These two very different government processes hold untold opportunities for different interpretations. Further, the legislative body commonly ignores the details of implementation, believing that, given some general principles, the executive branch will carry out the program as it was intended. In agriculture, Congress grants remarkable discretion to the Secretary of Agriculture who, in turn, goes to his agencies for information and advice. But agencies within government, such as the Agricultural Stabilization and Conservation Service within the U.S. Department of Agriculture, respond to a very different set of incentives than do elected officials. Agencies follow internal incentives that may or may not be appropriate for the larger goals of society. This leads to charges that bureaucracies are out of control both in their actions and in their spending. In part, those charges may be due to the agencies' internal incentives. Employees within government agencies are commonly rewarded for ideas that may justify the agency budget or lead to its expansion. Such activity is clearly inefficient.

Experience with a great variety of programs implemented in agriculture points to potential problems. Those held responsible for implementing or regulating a program are generally closely related to those they are trying to regulate. Meat inspectors were found to have close ties with managers of plants being inspected, which was considered part of the problem in USDA's meat inspection procedures during the 1970s. By way of solution, inspectors were required to move to different plants every two years to keep them from getting too cozy with plant managers.

The Agricultural Stabilization and Conservation Service (ASCS) is responsible for administering the price- and income-support programs for major agricultural commodities. At the county level, farmer members serve on the regulatory board. Quite naturally, such a system has problems because members find it difficult to impose sanctions or penalties on neighbors who may have violated some ASCS procedure. Local committee chairs of ASCS offices have even sent letters to their farmers suggesting ways to reduce the damage of acreage set-aside programs, for example by setting aside poor land or land that is isolated and hard to reach. Such suggestions run counter to the stated objective of set-aside policies to reduce production.

Because bureaucrats are motivated by job security, they sometimes extend and even create work in some problem areas even after producers or society no longer have a need. Job security behavior can also discourage innovation because creative ideas often carry a certain risk. Thus governmental inefficiencies can occur in the gap between what is possible and what is practical.

Another serious implementation problem arises when farmers know more than bureaucrats do. Consider the example of crop yields used to provide crop insurance coverage. In this case, farmers know more about their own yields than the government or any insurance company can ever expect to learn. Farmers but not the other party to any deal will know when a good or a bad offer is being made and will decide whether to participate on the basis of this knowledge. This situation creates serious financial difficulties for federal crop insurance. Those farmers who know how to work the system will be the real gainers from almost any public policy that is ostensibly designed to help all farmers.

*2. The Unintended Side Effect Problem.* Another failure of government involves the unintended side effects of some government activity. Like externalities in the market, these effects are difficult to predict and control.[1] Policymakers, particularly elected officials, live in the short run. They are seldom rewarded for considering long-run implications of the policies they enact. Especially when a problem is portrayed as a crisis, politicians can be caught up by strong political forces and can make decisions before the side effects can be considered. Agricultural policy is

*Policymakers, particularly elected officials, live in the short run. They are seldom rewarded for considering long-run implications of the policies they enact.*

Box 10.2.

## The Size and Weight of Agricultural Policy

The preamble to the 1990 farm bill, "Purpose of the Act," reads as follows: "To extend and revise agricultural price support and related programs, to provide for agricultural export, resource conservation, farm credit, and agricultural research and related programs, to ensure consumers an abundance of food and fiber at reasonable prices, and for other purposes."

That Act gives new meaning to the old concept of an omnibus bill—or, as punsters often joke, "ominous legislation." The 1990 farm bill has 25 titles, the first 10 of them for basic commodities. Each title is made up of numerous provisions. Cotton, for example, walks us separately through such provisions as suspension of base acreage, sales price restrictions, nonapplicability of the Agricultural Act of 1949, skip-row practices, and assorted other concerns. So much for the idea that people worry only about loan rates, target prices, and disaster payments.

The last 15 titles of the Act show even more clearly how to get lost in the agricultural policy jungle. Title 14 takes us to conservation, title 15 to trade, title 17 to food stamps, title 19 to product promotion, title 21 to organic certification, and title 24 to global climate change. Lest anything be forgotten, title 25 is for "other provisions." Look them up at your leisure!

full of examples. Commodity price supports designed to aid farmers cause land prices to increase, an effect that hurts those producing agricultural commodities on recently purchased land. Price supports may also damage or eliminate international markets, an effect that over time may well cause farm incomes to be lower than they may have been without price supports.

Another example of the unintended side effects of commodity programs can be seen in the incentives they created for farming marginal and erosive soils. Although erosion control is a goal of U.S. policy, the design of major commodity programs has actually encouraged farmers to plant row crops in areas that are better suited to pasture. There was no intent to do this; farmers simply planted to maximize their incomes under allowable practices. Trade-offs such as this are the rule in policy, not the exception. For nearly every program, we could point to positive and negative consequences. The trick is to anticipate beforehand when a program will have unintended side effects.

3. *The Size and Weight Problem: Too Big.* Modern agricultural policy has spawned a vast number of programs and regulatory authorities. The effective and timely oversight of them is therefore

impossible in even the most decentralized Congress and Administration (see Box 10.2). Change, after all, requires some concurrence from both branches of government and must be consistent with the federal budget. The potential loss of appropriations, agency jobs, services expected by constituents, and perhaps even congressional committee control is a Pandora's box few approach with relish. Too often, even when faced with obvious irrelevance, policymakers can avoid making changes because they can attend to so many other less risky policy matters.

The politically neglected Animal and Plant Health Inspection Service (APHIS) is a good case in point. While Congress has been studying the need for new guidelines for the contentious issues of biotechnology products, the less trendy work done by APHIS on animal disease control and eradication draws mostly yawns. In many ways, agency personnel like things that way. Because Congress is inattentive and key livestock interests take a proprietary view of animal health programs, APHIS is seldom pressured to change its ways.

Yet agency officials are aware that neglect breeds public policy atrophy. Agency personnel have resisted change despite evidence of increasing worldwide animal disease problems. So while many APHIS policy analysts have urged the agency to move away from traditional disease eradication and redirect its resources to disease detection and management, resistance within the agency has been substantial. APHIS's administrative management team, in fact, knows that the major diseases for which it is funded have almost been eradicated. But now another problem of size and weight arises. How, in the face of a job nearly done, does the organization move on officially to emphasize disease detection and management? It doesn't, at least not very well or efficiently. Not surprisingly, APHIS keeps a low profile in Washington politics, and Congress tries hard to ignore it. Everyone is left hoping that APHIS—like so many other parts of agricultural policy—can just muddle along without fostering a crisis that demands broader attention.

*4. The Fairness Problem: Everyone Should Share.* Because of government involvement, the distribution of wealth may be more inequitable than it would be without a particular public policy. For example, because price supports cause land prices to increase, landowners gain wealth. If they are already wealthy, then this type of government-induced wealth may be contrary to what we as a society desire regarding equity. In a democratic society, political power influences the actions of government. Political power may be distributed in such a fashion that wealthy people can become wealthier through government action, even if the larger society believes that the current distribution of wealth is unjust.

The size and weight problem is so severe, in part, because the policy process in U.S. agriculture creates continuing inequity. Correcting inequity is a consuming task. Presenting the appearance of correcting it and then advertising the wonderful result is also consuming. The more laws that are passed for specific beneficiaries, the greater the potential for program spillover effect. The more spillover, the more likely it is that third parties are harmed by the contract between the intended beneficiary of the first agreement and government as provider. The more that third parties are harmed, the more new programs develop as a corrective measure. The more new programs that are created, the more government must work to justify them. The cycle is vicious. The reason for it is clear. No member of Congress wants to go back home and be reminded that a government program results in state or district unpleasantries.

Countless examples illustrate this point. In the 1985 farm bill deliberations, international market forces determined the need for lowering loan rates. But with the high-blown rhetoric of family farm protection, few public officials wanted to be seen as unfair to farm program recipients. So, despite the preoccupation of Congress with the national budget deficit that year, target prices were kept relatively stable. Better that program costs should skyrocket than that Congress and the Administration should take away without giving fair return.

Farm bills are frequently rewritten solely to preserve the appearance of fair return. As representative Jamie L. Whitten (D-MS) once said, "I've never seen a disaster that wasn't an opportunity" (Rapp, p. 83). Provisions of the 1985 Act were revisited three times in 1986. No one flinched, at least in public, when a freshman representative discovered that dry-bean growers were hurt by flexible planting allowances granted to program crop producers. Rather, Congress wrote in legislative protection, limiting the originally intended flexibility. Much to the relief of all legislators, this rush to fairness opened the door to protecting others. Even the handful of southwest farmers who grow specialty crops for a natural laxative gained an unbound legislative reprieve. To district farmers, several legislative heroes were discovered as the stories of program corrections were repeated. Of course, these "corrections" defeated the original purpose of flexible planting.

> *Government is not only unfair if it does for one and harms another, but it is no less a villain if it does for one without doing for another.*

By opening agricultural policy to the poor, consumers, environmentalists, animal rights activists, and organic farmers, Congress has faced even more severe fairness problems with slight variations on the traditional theme. Government is not only unfair if it does for one and *harms* another, but it is no less a villain if it does for one *without doing* for another.

Two recent 1990 cases stand out. Facing unhappy food processors at home, many legislators wanted to forget more stringent food labeling laws. But without first dealing with the well-advertised health needs of consumers, how could Congress fairly consider a farm bill? The same was true of rural development. Until the very end of conference committee mark-up, even after a separate rural development bill died of its own slow pace and lack of focus, weary legislative advocates argued that the 1990 session could not end if Congress acted again for farmers and forgot other rural residents. With that rebuke, Senate and House conferees allowed fifteen minutes for antagonists to work out their differences and secure a few provisions.

Over and over, Congress feels compelled to give constituents the good news that Jimmy Carter was wrong about the world not being a fair place. Unfortunately, the explanations shed very little light on exactly what it was that programs were intended to produce in the first place. Overselling what has been done for every special constituency creates false expectations and, especially later, an endless search for data to justify the original policy deed.

5. *The Public Problem: Everyone Wins.* Perhaps the most repeated yet spurious comment made about farm price-support programs is that U.S. consumers pay amazingly little for their food, less than 12 percent of their 1990 income. Cause and effect, so hard to prove or disprove, nonetheless are easily lumped together in appealing to a popular image for agricultural policy. Government compulsively tries to show that the narrowest public policy benefit going to the most select farm or business constituent actually serves the broader public good. With so much emphasis on fairness among parties and with continually burgeoning programs, the need for such excuses is understandable. Few even sympathetic agricultural observers are prepared to swallow just how much, in several categories, the United States does for so few.

The rhetoric about program intent can get incredibly thick on such issues. Try, with a straight face, telling knowledgeable consumers that they too benefit when government buys out the herds of several dairy producers to lower supplies and keep milk prices high. Or try telling environmentally concerned Americans that the Conservation Reserve Program is really an environmental program rather than a supply-control effort when several other alternatives bring more conservation with far fewer dollars.

The myth of the average family farm and the near total disregard of macroeconomic policy effects on farm financial stress, as seen in earlier chapters, are two examples of even greater obfuscation in policy rhetoric. Government can hardly provide any form of direct assistance that benefits all categories of family farms. Yet both Congress and USDA continue to act as though most of their policies help all farmers.

Box 10.3.

## The Woeful Trap: "I Know a Farmer Back Home"

The day had dragged on, as members of Congress, debating the passage of major legislation, discussed whether to cut back on transfer payments to operators of large farms and to target payments to smaller farms. Calls from the district had been pouring in all day, several of them from farmers who were watching the floor debate on television. One farmer who called carefully explained, yet again to a legislator he knew well, why he was dependent on farm payments.

His story was simple. Without his current deficiency payment, he wouldn't be able to make loan payments on land he had purchased some years before when the payment program was expected to continue. If Congress decided to limit those payments, he would be partly cut off because his farm income was too high.

After the Congressman hung up the phone, he called in his legislative director. "Frank just called," he said. "He's afraid that what we're doing will cause him to lose his farm. We can't let this amendment go through. It's not right for us to change policy and see this good farmer's efforts go down the drain. He doesn't deserve that kind of treatment." The amendment to target payments away from large farmers lost.

No doubt other members of Congress took part in a similar scenario. Real people and real stories have a heavy influence. The plea is familiar: we should support farmers who base their major decisions on an expected stream of income from government payments. The farmer in this case clearly had spent more on land than he would have if there had been no deficiency payments at the time of purchase. But now he was trapped. The member of Congress was trapped too. So the status quo prevailed, guaranteeing that other farmers would fall into the same trap and buy land with the expectation that the payment program would continue.

Until we acknowledge that things get done simply because farmers want them done, public officials will go on claiming that dumping large sums of money on the largest producers is good for everyone. And the hectic search for any data that will suggest a link between the farm good and the consumer good will continue unabated. Of course, when the resulting information in support of the public interest is processed, not a single data user will be any the wiser about what agricultural programs are doing.

*6. The Omnipotence Problem: What We've Done Can't Be Wrong.*
The final reason why rhetoric disguises what programs do is more detrimental to sound agricultural policy results than are the other five reasons. After all, it seems somewhat laudable that public officials try to (1) do too much, as opposed to too little; (2) do something for everyone, as opposed to systematic neglect of a few; and (3) ascertain that many (as opposed to only single) interests benefit from each decision. An inability to equate program intent with the reality of what goes on means that at least someone knows enough to be trying. So we can take some small comfort in knowing that agricultural policy is not suffering from benign neglect.

Policymakers do, however, fail to identify and eliminate programs that may actually harm U.S. agriculture, its dependents, or its constituents. Because programs must be constantly justified, public officials spend most of their time and effort assembling data to support their position. In the end, they are left precious little time for rooting out disastrous effects. Those who contribute to program development have an omnipotence problem: they believe that what their political maneuvering has created cannot possibly go wrong.

Yet too many obvious failures exist for U.S. farmers and the general public to believe that the creators of agricultural policies were either all-powerful or all-knowing. These policymakers have created stakeholders who, as beneficiaries of otherwise failing programs, will not let go of the past, even if the past never contributed to positive results for most (see Box 10.3).

## Living with Irony

The contradictory myths that frame our expectations about the policy process are not likely to go away. There are too many flaws in how decisions are made to expect public satisfaction. Planning is necessary, even though it will necessarily fail. Those with long experience in government have learned to live with irony—to be guided by divided loyalties and inconsistent ideals. But irony is not a bad guide for policy, for it is a way of acknowledging that each of the myths guiding our policies contains a grain of truth. We emphasize this point to remind ourselves that no single interest, including our own, represents the public good.

The danger is that irony can degenerate into cynicism. The cynic sees *only* confusion in the morass of policy and forgets the need for good intentions. In practical terms, the cynic sees policy *merely* as a contest of special interests and nothing more. The cynic abandons all attempts to seek a common good. But the impulse toward cynicism must be resisted. One key step to purging cynicism is to welcome new interests that want to be accommodated in agricultural policy. Another is to make certain that traditional interests with legitimate needs do not get left out of the process. Pursuit of the common good starts with inclusive debate.

## Notes

1. Caution is advised when considering "unintended" consequences. It may well be that some members of Congress know the side effects of certain policies and know that they will benefit the "right" constituents. Therefore, what may be later identified as an unintended side effect may have actually been intended.

# 11 Conclusions

In this book, we have examined how beliefs about American rural life establish collective values of citizenship, community, and moral character for Americans. The Jeffersonian farmer understood that personal interest and the common good are one and the same. The Emersonian farmer developed the virtues of self-reliance and stewardship as a natural outgrowth of production practices. Our children continue to learn an appreciation for nature, hard work, and personal responsibility through agrarian images and stories. These messages are important for the collective American understanding of ethics and democracy. But it is not clear that the modern farm is better suited than the suburban home or office building to the development of values historically regarded as agrarian.

Our unexamined traditional agrarian beliefs play havoc with the agricultural policy process. Because Americans rightly want to preserve the values expressed in agrarian imagery, they resist attacks on agrarian myths. Indeed, in calling them myths, we acknowledge the enduring truth of the values that these beliefs transmit, but not every statement or application of agrarian values is sacred. Our desire to protect those mythic, self-reliant stewards of democracy that are part of our constitutional foundation has led Americans to enact and to persist in policies that have contradictory and destructive consequences. Each individual must be a self-reliant steward of democracy, not farmers alone (or the institution of agriculture). That means that Americans must now rethink the terms of debate for an agricultural policy dedicated to the common good.

Although it is no doubt clear that we, the authors, have some specific goals for change in agricultural policy, our primary objective has been to shift the burden of proof in agricultural policy disputes. Until now, that burden of proof has relied upon an inappropriate application of agrarian myth in at least five principal ways:

- Policymakers have assumed that policies that benefit farming benefit rural America as a whole.
- Policymakers have assumed that policies intended to aid an archetypal Everyman farmer would distribute assistance to farmers fairly, without regard to farm size, location, or economic and political power.
- Policymakers have assumed that because farmers produce food, anything that increases the production of food will help farmers.
- Policymakers have assumed that self-reliant farmers should focus primarily upon meeting national food needs.
- Policymakers have assumed that good farmers would be natural stewards of the entire rural environment.

Congress and decision makers have not made these assumptions uncritically, but they have required a strong burden of proof from people who introduce contrary statements into the policy debate. We have argued that, while these assumptions are not strictly false (they are, indeed, true in some cases), the burden of proof should be just the reverse. Policymakers should require those who advocate such agrarian claims to demonstrate conclusively that there are specific reasons for accepting those claims in special cases. In this manner, debate over agricultural policy may free itself from the inertia that has plagued it in past decades.

We present the following guidelines for those interested in reform:

- Develop specific programs to address rural development needs. While many of these needs are more appropriately the concern of state and local governments, the federal government has a role in providing services to rural areas. Rural development can be farm policy if it broadens opportunities for small-scale farmers.
- Recognize that farm size, location, and economic circumstances influence who gains and who loses from federal agricultural programs. If benefits continue, they must be directed to particular classes of farms if they are to achieve any meaningful social or economic goals.
- Insist that farm program benefits, if they continue, no longer be tied to production. By linking program outlays with production, policies inflate land prices and make U.S. agriculture less competitive in international markets. Furthermore, production-linked benefits tend to encourage more environmental damage than would otherwise occur.

- Recognize that an uncertain macroeconomic environment will play a major role in determining farm income, but that these uncertainties cannot be addressed through specific agricultural policies.
- Recognize the need for institutions that can help farmers manage risk in a way that does not influence land prices or production.
- Recognize that there are conflicting objectives in existing agricultural programs that attempt to address environmental concerns. New programs must be goal directed, not means directed.
- Determine food programs for the hungry and disadvantaged on the basis of need, not on the basis of farm program goals.

This agenda is ambitious, but unless we go back to first principles, U.S. food and agricultural policy will become increasingly irrelevant to all but a few narrow interests. While public officials and members of the public at large may accept agricultural policies because they are hard to change, little about them is right.

Few Americans are satisfied with agricultural policies, and fewer still can find arguments powerful enough to motivate meaningful change. We began this book under the assumption that fealty to agrarian values was becoming part of the problem, rather than part of the solution. We end it by repeating our statement that what is right about agrarian values is a large part of what is right about American democracy. These values must be preserved and nurtured. But what is wrong in the application of agrarian values is a large part of what is wrong with agricultural policy. The application of those values to policy must be rethought, refocused, and reformed.

We hope that our book is less a debunking of agrarian myth than a *rebunking*. We Americans must preserve our traditions, but must not allow them to become excuses for weakness and special interests. There is a common good in agriculture. Agrarian myths can help us find that good if we apply the myths critically. We have not specified policy options in detail, for the common good can only manifest itself through open-ended political debate. We have aimed to reorient that debate, but its future outcome is truly undetermined. We thank our readers who will carry on the debate and move it to its next stages.

# Bibliography

Ahearn, Mary, Gerald Whittaker, and Hisham El-Osta. In press. "The Production Cost-Size Relationship: Measurement Issues and Estimates for Three Major Crops." In *Size and Structure Issues in American Agriculture*, edited by Arne Hallam. Boulder, Colo.: Westview Press.

Barton, Weldon V. 1976. "Coalition-Building in the U.S. House of Representatives: Agriculture Legislation." In *Cases in Public Policy*, edited by James E. Anderson, 141–162. New York: Praeger.

Bentham, Jeremy. 1970 [1789]. *An Introduction to the Principles of Morals and Legislation*, edited by J. H. Burns and H. L. A. Hart. London: Athlone Press.

Berry, Wendell. 1977. *The Unsettling of America*. San Francisco: Sierra Club Books.

———. 1981. *The Gift of Good Land*. San Francisco: North Point Press.

Bonnen, James T., and William P. Browne. 1989. "Why is Agricultural Policy So Difficult to Reform?" In *The Political Economy of U.S. Agriculture*, edited by Carol S. Kramer, 7–15. Washington, D.C.: Resources for the Future.

Bovard, James. 1989. *The Farm Fiasco*. San Francisco: Institute for Contemporary Studies Press.

Breimyer, Harold F. 1965. *Individual Freedom and the Economic Organization of Agriculture*. Urbana: University of Illinois Press.

———. 1977. *Farm Policy: 13 Essays*. Ames: Iowa State University Press.

Brewster, John M. 1963. "The Relevance of the Jeffersonian Dream Today." In *Land Use Problems and Policy in the United States*, edited by H. W. Ottoson, 86–136. Lincoln: University of Nebraska Press.

Browne, William P. 1988. *Private Interests, Public Policy, and American Agriculture*. Lawrence: University Press of Kansas.

———. 1990. "Organized Interests and Their Issue Niches: A Search for Pluralism in a Policy Domain." *Journal of Politics* 52:477–509.

Burke, Edmund. 1968 [1790]. *Reflections on the Revolution in France*, edited by C. C. O'Brien. Harmondsworth, U.K.: Penguin Books.

Buttel, Frederick H., and Louis E. Swanson. 1986. "A Farm Structural and a Public Policy Context of Soil and Water Conservation." In *Conserving Soil: Insights from Socioeconomic Research*, edited by Stephen B. Lovejoy and Ted L. Napier, 26–39. Ankeny, Iowa: Soil Conservation Society of America.

Cochrane, Willard W. 1979. *The Development of American Agriculture: A Historical Analysis*. Minneapolis: University of Minnesota Press.

Congressional Budget Office, U.S. Congress. 1977, January. *The Food Stamp Program: Income or Food Supplementation?* Washington D.C.: U.S. Government Printing Office.

Corrington, Robert S. 1990. "Emerson and the Agricultural Midworld." *Agriculture and Human Values* 7(1):20–26.

Cowan, R. S. 1983. *More Work for Mother*. New York: Basic Books.

Crosson, Pierre. 1986. "Soil Conservation: It's Not the Farmers Who Are Most Affected by Erosion." *Choices* 1:33–38.

Crutchfield, Stephen R. 1989, November. "Agriculture's Effects on Water Quality." In *Agricultural-Food Policy Review: U.S. Agricultural Policies in a Changing World*, 369–382. Agricultural Economic Report No. 620. Washington, D.C.: Economic Research Service, U.S. Department of Agriculture.

Deavers, Kenneth. 1990. "Rural Vision—Rural Reality: Efficiency, Equity, Public Goods, and the Future of Rural Policy." Benjamin H. Hibbard Memorial Lecture Series, Department of Agricultural Economics, College of Agricultural and Life Sciences, University of Wisconsin–Madison, April 20, 1990.

Dunham, Denis. 1991, March. *Food Costs . . . From Farm to Retail in 1990*. Agriculture Information Bulletin, No. 619. Washington, D.C.: Economic Research Service, U.S. Department of Agriculture.

Emerson, Ralph Waldo. 1965. *Selected Writings of Ralph Waldo Emerson*, edited by William H. Gilman. Ontario: The New American Library.

Espelin, Arnold L., Arthur H. Grube, and Virginia Kibler. 1991, July. *Pesticide Industry Sales and Usage 1989 Market Estimates*. Washington, D.C.: Economic Analysis Branch, Biological and Economic Analysis Division, Office of Pesticide Programs, EPA.

French, B. 1982. "Fruit and Vegetable Marketing Orders: A Critique of the Issues and State of Analysis." *American Journal of Agricultural Economics* 64:916–923.

Gardner, Bruce L. 1981. *The Governing of Agriculture*. Lawrence: University of Kansas Press.

Goldschmidt, Walter. 1978. *As You Sow: Three Studies in the Social Consequences of Agribusiness*. Montclair, N.J.: Allanheld, Osmun & Co.

Goss, Kevin F., Richard D. Rodefeld, and Frederick H. Buttel. 1980. "The Political Economy of Class Structure in U.S. Agriculture: A Theoretical Outline." In *The Rural Sociology of the Advanced Societies*, edited by Frederick H. Buttel and Howard Newby, 83–132. Montclair, New Jersey: Allanheld, Osmun & Co.

Hardwood, Richard. 1991. *Citizens and Politics: A View from Main Street America*. New York: The Kettering Foundation.

Harrison, Welford. 1972. "The Forgotten Man: The Federal Meat Inspector." In *Sowing the Winds*, edited by Welford Harrison, 45–78. New York: Grossman.

Hathaway, Dale E. 1963. *Government and Agriculture*. New York: Macmillan.

Heady, Earl, and Carl Allen. 1951. *Returns from Capital Required for Soil Conservation Farming Systems*. Research Bulletin 381. Ames: Experiment Station, College of Agriculture, Iowa State University.

Hightower, Jim. 1972. *Hard Tomatoes, Hard Times: The Failure of the Land Grant College Complex*. Cambridge: Schenkman Publishing.

Houck, J. 1990. "Stabilization in Agriculture: An Uncertain Quest." In *Agricultural Policies in a New Decade*, edited by K. Allen. Washington, D.C.: Resources for the Future.

Jefferson, Thomas. 1984. *Writings*, edited by Merrill D. Peterson. New York: Literary Classics of the United States.

Kramer, C. 1990. "Impacts of the 1990 Farm Bill on Consumers." Paper presented at the winter meetings of the American Agricultural Economics Association, December 1990.

MacCannell, Dean. 1988. "Industrial Agriculture and Rural Community Degradation." In *Agriculture and Community Change in the U.S.: The Congressional Research Reports*, edited by L. E. Swanson, chap. 2. Boulder, Colo.: Westview Press.

Mercier, S. 1989, September. "Corn: Background for 1990 Legislation." Staff Report No. 89-47. Washington, D.C.: Economic Research Service, U.S. Department of Agriculture.

Paarlberg, Don. 1980. *Farm and Food Policy: Issues for the 1980s*. Lincoln: University of Nebraska Press.

Paarlberg, R. L. 1990. "The Mysterious Popularity of EEP." *Choices* 5(2):14–17.

Rapp, David. 1988. *How the U.S. Got Into Agriculture: And Why It Can't Get Out*. Washington, D.C.: Congressional Quarterly.

Reichelderfer, Katherine. 1990. "Environmental Protection and Agricultural Support: Are Tradeoffs Necessary?" In *Agricultural Policies in a New Decade*, edited by Kristen Allen, 201–230. Washington, D.C.: Resources for the Future.

Reichelderfer, Katherine, and Maureen Kuwano Hinkle. 1989. "The Evolution of Pesticide Policy: Paving the Way for Environmental Interests and Agriculture." In *The Political Economy of U.S. Agriculture: Challenges for the 1990s*, edited by Carol S. Kramer, 147–173. Washington, D.C.: Resources for the Future.

Reid, Norman. 1989, December. *Rural America: Economic Performance, 1989*. Washington, D.C.: Agriculture and Rural Economy Division, Economic Research Service, U.S. Department of Agriculture.

Reimund, Donn, and Nora Brooks. 1990. "The Structure and Status of the Farm Sector." In *The U.S. Farming Sector Entering the 1990's: Twelfth Annual Report on the Status of Family Farming*, 7–15. Agricultural Information Bulletin, No. 587. Washington, D.C.: Economic Research Service, U.S. Department of Agriculture.

Ribaudo, Marc O. 1986, September. *Reducing Soil Erosion: Offsite Benefits*. Agricultural Economic Report No. 561. Washington, D.C.: Economic Research Service, U.S. Department of Agriculture.

Scott, John T., Jr. 1989, December. *Lease Shares and Farm Returns*. AE-4657. Urbana: University of Illinois, Department of Agricultural Economics.

Shaffer, James Duncan. 1990. "The Distribution of Direct Payments to Farm Operators in 1986 and 1987: Some Questions about Policy Objectives." Discussion Paper Series, No. FAP90-08. Washington, D.C.: Resources for the Future.

Skees, Jerry R., and Louis E. Swanson. 1988. Farm Structure and Rural Well-Being in the South." In *Agriculture and Community Change in the U.S.: The Congressional Research Reports*, edited by L. E. Swanson, chap. 6. Boulder, Colo.: Westview Press.

Strange, Marty. 1988. *Family Farming: A New Economic Vision*. Lincoln: University of Nebraska Press.

Sugarman, Carole. 1991, June 5. "Catering to Cows and Consumers: Is the USDA Caught in a Conflict of Interest?" *The Washington Post,* E1–3.

Swanson, Louis E. 1989, Summer. "The Rural Development Dilemma," *Resources* 96:14–17.

Thompson, Paul B. 1988. "The Philosophical Rationale for U.S. Agricultural Policy." In *U.S. Agriculture in a Global Setting,* edited by M. Ann Tutwiler, 34–45. Washington, D.C.: Resources for the Future.

Thoreau, Henry David. 1970 [1854]. *Walden, or Life in the Woods,* edited by Philip VanDoren Stern. New York: Crown Publishers.

Tiner, R. W., Jr. 1984. *Wetlands of the United States: Current Status and Recent Trends.* U.S. Department of the Interior, Fish and Wildlife Service. Washington, D.C.: U.S. Government Printing Office.

Tyers, R., and K. Anderson. 1988. "Liberalizing Agricultural Policies in the Uruguay Round: Effects on Trade and Welfare." *Journal of Agricultural Economics* 39:197–216.

U.S. Department of Agriculture. 1989. *Agricultural Statistics.* Washington, D.C.: U.S. Government Printing Office.

U.S. Department of Agriculture. 1990, August. "Multilateral Trade Reform: What the GATT Negotiations Mean to U.S. Agriculture." U.S. Department of Agriculture staff briefing.

U.S. Department of Agriculture. 1991, February. *World Grain Situation and Outlook.* FG2-91. Washington, D.C.: U.S. Government Printing Office.

U.S. Department of Agriculture, Economic Research Service. 1989, September. *Agricultural Resources Situation and Outlook.* Washington, D.C.: Economic Research Service, U.S. Department of Agriculture.

U.S. Department of Agriculture, Economic Research Service. 1990, April. *Economic Indicators of the Farm Sector: Cost of Production, Major Field Crops, 1988.* Washington, D.C.: Economic Research Service, U.S. Department of Agriculture.

U.S. Department of Agriculture, Economic Research Service. 1990. *Economic Indicators of the Farm Sector: National Financial Summary.* Washington, D.C.: Economic Research Service, U.S. Department of Agriculture.

U.S. Department of Agriculture, Economic Research Service. 1991, February. *Economic Indicators of the Farm Sector: State Financial Summary, 1989.* Washington, D.C.: Economic Research Service, U.S. Department of Agriculture.

U.S. Department of Commerce, Bureau of the Census. 1990. *Statistical Abstract of the United States.* Washington, D.C.: U.S. Government Printing Office.

Webb, Alan J., Michael Lopez, and Renata Penn, eds. 1990. *Estimates of Producer and Consumer Subsidy Equivalents: Government Intervention in Agriculture, 1982–87.* Statistical Bulletin No. 803. Washington, D.C.: Agriculture and Trade Analysis Division, Economic Research Service, U.S. Department of Agriculture.

World Bank. 1988. *World Development Report.* New York: Oxford University Press.

# About the Authors

**William P. Browne:** Ph.D. Washington University, St. Louis; political scientist, professor and director of Public Administration Programs at Central Michigan University. Author of *Private Interests, Public Policy and American Agriculture* (1988), Browne has focused most of his research for the past fifteen years on the agricultural policy process. In addition to more than sixty publications, he has coauthored or coedited six other books on facets of agriculture and rural policy. As a visiting fellow with the Economic Research Service (1985) and the National Center for Food and Agricultural Policy (1991), he spent considerable time on Capitol Hill, with lobbyists from throughout the agriculture sector, and at the grass-roots with U.S. farmers.

**Jerry R. Skees:** Ph.D. Michigan State University; agricultural economist, professor and director of Graduate Studies, Department of Agricultural Economics, University of Kentucky. Skees is author or coauthor of more than fifty publications in the area of agricultural policy and rural development. He has been active in working with farmers, agribusiness leaders, and state and national policymakers. In 1989, Skees served as research director for the Federal Commission on Crop Insurance and as visiting scholar in the Economic Research Service of the U.S. Department of Agriculture. During that time he also served on the Secretary of Agriculture's Task Force on Disaster Assistance policy for the 1990 farm bill.

**Louis E. Swanson:** Ph.D. Pennsylvania State University; sociologist, associate professor, Department of Sociology, University of Kentucky, Lexington. His research has focused on the sociology of agriculture and change in the rural community, with an emphasis on agricultural and rural development policy. In addition to forty-one articles and book chapters, Swanson has edited or coedited three books on these subjects, *Agriculture and Community Change in the U.S.*, *American Rural Communities*, and *Farming and the Environment*, and has coauthored *Rural Communities*, a forthcoming textbook on rural social problems. He has also contributed to two Congressional Office of Technology Assessment studies and has testified before Congress on several occasions on agriculture and rural community issues. Swanson was a visiting scholar with the National Center for Food and Agricultural Policy, 1988 and 1989.

**Paul B. Thompson:** Ph.D. in philosophy from the State University of New York at Stony Brook; director of the Center of Biotechnology Policy and Ethics and associate professor of philosophy and of agricultural economics at Texas A&M University, where he also holds the Maria Julia and George R. Jordan, Jr., Professorship in Public Policy. Thompson has served as president of the Agriculture, Food and Human Values Association and was director of Graduate Studies in Philosophy during the inaugural two years of the M.A. program in philosophy at Texas A&M University. He has been a fellow at the Council on Foreign Relations and at the National Center for Food and Agricultural Policy. His current projects center on the analysis and communication of risk, particularly as it relates to food, agriculture, and the transfer of genetic materials. He is coeditor with Bill Stout of *Beyond the Large Farm: Ethics and Research Goals for Agriculture*, published by Westview Press in 1991, and author of *The Ethics of Aid and Trade* (forthcoming).

**Laurian J. Unnevehr:** Ph.D. Stanford University; agricultural economist, associate professor, Department of Agricultural Economics, University of Illinois at Urbana–Champaign. From 1982 to 1985, she was a Rockefeller Foundation social science postdoctoral fellow at the International Rice Research Institute in the Philippines. Her research, reported in more than twenty journal articles and book chapters, has encompassed the food policy issues associated with international trade, marketing, and consumer demand. As part of a project funded by the W. K. Kellogg Foundation's program for innovative policy education, Unnevehr organized a series of roundtables where farm and nonfarm interest groups in Illinois discussed policy issues for the 1990 farm bill. At these roundtables, she witnessed firsthand the impact of agrarian myths on public perceptions and policy dialogue.

# About the Book

*Sacred Cows and Hot Potatoes* challenges many of the assumptions of current agricultural policies—such as equating "farm" with "rural," high farm prices with high farm incomes, or farm programs with food programs—and examines the agrarian roots of these policies. From the origins of agrarian myths to the latest controversies over farming and the environment, this book provides an overview of the use and abuse of agrarian values in policymaking. Illustrated with pictures, cartoons, and graphs, the book will appeal to a broad audience, including policymakers, rural sociologists, agricultural economists, political scientists, ethicists, and the interested public.

Printed and bound by CPI Group (UK) Ltd, Croydon, CR0 4YY

23/10/2024

01778259-0018